Outlines in Pharmaceutical Sciences

T0075769

For further volumes:
http://www.springer.com/series/8870

Anthony J. Hickey · Hugh D.C. Smyth

Pharmaco-Complexity

Non-Linear Phenomena and Drug Product Development

 Springer

Anthony J. Hickey, Ph.D.
University of North Carolina
Eshelman School of Pharmacy
Chapel Hill, NC 27599, USA
ahickey@unc.edu

Hugh D.C. Smyth, Ph.D.
University of Texas
Austin, TX 78712, USA
hsmyth@mail.utexas.edu

ISSN 2191-3188   e-ISSN 2191-3196
ISBN 978-1-4419-7855-4   e-ISBN 978-1-4419-7856-1
DOI 10.1007/978-1-4419-7856-1
Springer New York Dordrecht Heidelberg London

Printed on acid-free paper

Springer is part of Springer Science+Business Media (www.springer.com)

Preface

The interpretation of physical, chemical, and biological phenomena as linear relationships between variables, or as simple functions of the variables, has been a significant scientific and mathematical strategy to their elucidation for centuries. It is often the case that the nature of linearity is to follow mathematical functions, e.g., power, exponential, or logarithmic functions, nevertheless the desire to fit data to simple predictable expressions is imbued in every scientist and engineer. From a philosophical standpoint, there is no reason to criticize this approach as it allows us to interpret the natural world and has a lofty heritage going back to the classical world.

However, nonlinear phenomena have been identified in many fields and interpreted as periodic, catastrophic, chaotic, or complex involving a variety of mathematical tools for analysis. Benoit Mandelbrot's classic book on the fractal geometry of nature and the many subsequent texts, most recently Wolfram's magnum opus *A New Kind of Science*, have raised questions about the nature of reality and the interpretation of observed phenomena. It seems clear that the complexity of dynamic events (on any scale) can rarely be explained by linear interpretations. The rare exceptions are likely to represent a convergence of multiple phenomena giving the appearance of a linear relationship between variables.

In fields related to pharmaceutical sciences, some texts have been written by pioneers such as Brian Kaye. His eminently readable *A Random Walk Through Fractal Dimensions* and *Chaos and Complexity* were seminal volumes for the authors. Tracing the mathematics of complexity back to the nineteenth century and beyond gives validity to the search for more accurate interpretations of experimental observations that should impact on the pharmaceutical sciences as significantly as other fields of endeavor.

The chemistry and physics literature are replete with papers on complexity from such notables as Ilya Prigogine and Murray Gell-Mann. A broad range of biological phenomena, the most complex imaginable from molecular biology to ecology, is now the subject of complexity analysis. Pharmaceutical sciences encompass biology, chemistry, physics, and mathematics associated with drug discovery, delivery, disposition, and action. This text describes a range of topics of importance in the pharmaceutical sciences that indicate a need for a nonlinear interpretation if they are to be characterized accurately, understood fully, and potentially controlled or modulated in the service of improved therapeutic strategies.

It is likely that the future will involve increasingly complex interpretations of data related to drug design and delivery, particularly as our knowledge of the human genome leads inexorably to the potential for individualized therapy. We hope that this text will promote discussion of the varied phenomena leading to pharmacological effect and the complex interactions, ultimately resulting in improved disease control and health maintenance.

Chapel Hill, NC   Anthony J. Hickey
Austin, TX   Hugh D.C. Smyth

Contents

Chapter 1
The Nature of Complexity and Relevance to Pharmaceutical Sciences

The decade of the 1970s was a period of change for many reasons. In addition to the revolution in personal computing, information, and gaming technology, a new view of the underpinning mathematics and science in nature was evolving. Benoit Mandelbrot developed a mathematical approach to describe the apparent complexity in nature (Mandelbrot 1977). He observed that the superficial appearance of objects was built on a foundation of nonlinearity or "roughness" which could be described simply as self-similar at any scale of scrutiny. This "fractal" approach became a central philosophy behind complexity studies. As scientists in different disciplines began to accommodate complex interpretations of their data, a new approach to predict or describe physical phenomena evolved that challenged more traditional methods and increased the potential to understand previously poorly understood. During this period, others were also formulating approaches to nonlinear phenomena in a variety of fields (Woodcock and Davis 1978). James Gleick's popular book *Chaos* describes in detail the exciting story of the birth and development of the field of complexity in this period and the skepticism with which the original protagonists were greeted by the mainstream scientific community (Gleick 1987). Three decades after these developments, there is sufficient evidence for the nonlinearity of many natural phenomena to propose that we might begin to interpret similar observations in pharmaceutical sciences.

A.J. Hickey and H.D.C. Smyth, *Pharmaco-Complexity*, Outlines in Pharmaceutical Sciences 1, DOI 10.1007/978-1-4419-7856-1_1, © American Association of Pharmaceutical Scientists 2011

The major fields/topics within the pharmaceutical sciences are no less open to these new advances than any other area, but there has yet to be a treatment of these broad principles in a single text. This volume is intended to inform those who have an interest and to make accessible some of the underlying principles in the hopes that a new generation of researchers will gain insight into the complexity associated with pharmaceutical systems. The following brief overview attempts to identify important areas in which new mathematical approaches may prove useful in future data interpretation.

A range of phenomena have been identified that have multiple underlying mechanisms that may lead to simple or complex outcomes, which have historically been reduced to best fit mathematical interpretations that do not necessarily result in greater understanding. The data usually encodes much more information about the phenomenon being studied than can be derived from a simple best fit model. The challenge is to differentiate the information from random events in order to create knowledge about the system that can be used to promote understanding with implications for control and quality of experiments or products. To guide our thinking on this subject, Fig. 1.1a has been used to indicate the foundation of data that creates information suitable for knowledge of a system and hopefully creates the wisdom to predict and control

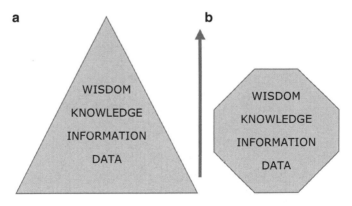

Fig. 1.1 Schematic depictions of information hierarchy as: (**a**) usually depicted and; (**b**) accurate reflection of reality

phenomena of importance. After a little consideration, it will occur to most scientists that we more frequently take a limited body of data or data from which information is being derived. Moreover, it is apparent that knowledge and wisdom derived from these data are perhaps not as illuminating as we would desire as shown in Fig. 1.1b. In both of these depictions, the conventional assumption is that inductive reasoning evolves the understanding required to manage the process or system under evaluation.

The following text has been divided into major areas of importance in the pharmaceutical sciences which can be related to the product development roadmap: beginning in the area of drug discovery, which includes chemistry, pharmacology, and biochemistry of the drug molecule and its biological target (receptor, enzyme, and antibody), extending through the dosage form considerations (solid state chemistry and process engineering) into disposition following preclinical or clinical administration to animals or humans, respectively, and finally to the importance of population biology to the therapeutic effect. These are example topics from within the range of items that must be considered in product development. Figure 1.2a illustrates the overlap between these fields. Figure 1.2b begins to accommodate some of the other elements of interactions such as the scale of the contribution

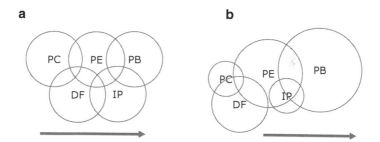

Fig. 1.2 Schematic diagrams of the relationship between elements of pharmaceutical development: (**a**) classical Venn diagram depicting perceived overlap/interaction; (**b**) extrapolation to capture magnitude of contribution of each. *PC* physico-chemical factors; *DF* dosage form; *PE* pharmaceutical engineering (manufacturing); *IP* individual pharmacokinetics and pharmacodynamics; *PB* population biology

(as indicated by the size of each circle). This is not intended to elicit discussion of the accuracy of the interpretation, rather it is to open up a discussion of whether we grasp the subtleties of the confounding factors and their contribution to our ultimate ability to understand, predict, and control important factors in product performance and therapeutic effect.

In conclusion, we will return to these themes and particularly the implicit directionality (indicated by arrows in the diagrams) of our thinking about our approach to the generation of knowledge and understanding of the influence of each discipline on others that appear to overlap.

References

Gleick J (1987) Chaos. Penguin Books, New York
Mandelbrot BB (1977) The fractal geometry of nature. W.H. Freeman, New York
Woodcock A, Davis M (1978) Catastrophe theory. Penguin Books, New York

Further Reading

Banchoff TF (1990) Beyond the third dimension. Scientific American Library, New York
Barnsley M (1985) Fractals everywhere. Academic, New York
Efros AL (1986) Physics and geometry of disorder. Mir Publishers, Moscow
Falconer K (1990) Fractal geometry, mathematical foundations and applications. Wiley, New York
Gilmore R (1981) Catastrophe theory for scientists and engineers. Wiley, New York
Kim JH, Stringer J (1992) Applied chaos. Wiley, New York
Liebovitch LS (1998) Fractals and chaos, simplified for the life sciences. Oxford University Press, Oxford
Moon FC (1992) Chaotic and fractal dynamics. Wiley, New York
Nayfeh AH, Balachandron B (1995) Applied nonlinear dynamics. Wiley, New York
Schroeder M (1991) Fractals, chaos, power laws. W.H. Freeman, New York
Stauffer D, Aharony A (1992) Introduction to percolation theory, 2nd edn. Taylor and Francis, Washington
Thompson JMT, Bishop SR (1994) Nonlinearity and chaos in engineering dynamics. Wiley, New York

Chapter 2
Phenomena in Physical and Surface Chemistry

Complex phenomena in pharmaceutical chemistry may have a significant impact on the performance of the dosage form. Traditionally, the concern has been with bulk product performance and stability with a focus on chemical degradation, and while this remains a serious issue over the last few decades, the focus on reproducible product performance has shifted toward considerations of physical stability.

A brief discussion of chemical considerations is followed by a more detailed presentation of the development of our understanding of important physical and surface chemical considerations.

Chemical Reactions

Certain chemical reactions are known to be complex phenomena. Molecular dynamics and chemical kinetics have been studied thoroughly, and they frequently appear predictable in terms of classical exponential or power functions (Billing and Mikkelsen 1996). The apparently predictable nature of chemical reactions may be an illusion of simplicity described by the Ergodic hypothesis (de Oliveira and Werlang 2007; Szasz 1994). Boltzmann originally conjectured that time-averaged behavior of microscopic components of a system gives the same outcome as the macroscopic, bulk average, where the bulk is a collection of all possible

A.J. Hickey and H.D.C. Smyth, *Pharmaco-Complexity*, Outlines in Pharmaceutical Sciences 1, DOI 10.1007/978-1-4419-7856-1_2, © American Association of Pharmaceutical Scientists 2011

states that molecules would reach in assembly in infinite time. Clearly depicting each of these states independently would be an enormously complex system. Attempts are being made to scrutinize reactions to more accurately depict the events that may be occurring (Nowak and Fic 2010; Bonchev et al. 1987).

Surface and Interfacial Chemistry

Adsorption constitutes an important area of research in which efforts to understand complex phenomena have extended for over a century. The following section begins by considering molecular association, extends to surface adsorption and models that have been proposed for nonlinear data fitting. The ability to measure surface features and energy densities at a molecular level has improved the potential to learn about the nature of interactions.

Surfactant molecules are noteworthy because of their capacity through discrete polar and nonpolar regions to align at interfaces, in particular the surfaces of solids in suspension.

The use of surfactants as coating materials requires consideration of the nature of these compounds, their interactions, and association with other substances.

Surface activity is a dynamic phenomenon, since the final state of a surface or interface represents a balance between the tendency towards adsorption and towards complete mixing due to the thermal motion of molecules.

A surface active agent (surfactant) may be described as a substance which alters the conditions prevailing at the interface. All surfactants are characterized by two structural regions, a hydrocarbon chain, which is hydrophobic, and a polar, hydrophilic group. The nature of the hydrophilic region of the surfactant enables the classification of surfactants to be subdivided into anionic, cationic, and nonionic. Examples of these are sodium dodecyl sulfate, dodecyl trimethyl ammonium bromide, and n-dodecyl hexaoxyethylene glycol monoether, respectively. Two further groups of surfactants exist: ampholytic surfactants which are zwitterionic and can behave as any of the aforementioned examples depending on the pH at which they are maintained, such as alkyl betaine, and natural surfactants which usually contain a glycerol moiety, such as phosphatidylcholine. Therefore, surfactants may be described as amphiphilic.

The formation of aggregates (Kertes 1977) and micelles (Eike 1977; Ravey et al. 1984) in solutions of surfactants is well documented. The term "micelle" should designate any soluble aggregate spontaneously and reversibly formed from amphiphilic molecules or ions (Tanford 1980). The micellization processes according to the commonly used equilibrium thermodynamic descriptions, namely the multiple equilibrium model and the pseudo-phase model, are, like the micelle definition, equally well applicable to aqueous and nonpolar solutions (Mukerjee 1974). The second model best conforms to the definition of a micelle described above.

The interactions governing the formation of surfactant aggregates in apolar media are different from those in aqueous solutions, in spite of the apparently similar building principle of lipophilic and hydrophilic micelles. The differences between interactions encountered in aqueous and nonpolar surfactant solutions have been considered at a molecular level with reference to the stability or existence of micelles in apolar media (Eike 1977). It has been concluded that once equilibrium between monomers and micelles, equivalent to the pseudo-phase model, ceases to be operative and is replaced by a stepwise aggregation equilibrium, the concept of a critical micelle concentration (CMC) is inapplicable. The two models for the process of interaction between surfactant molecules, described above, are, therefore, considered to be mutually exclusive (Kertes 1977).

Aggregation

Inverted or reversed micelles are examples of molecular aggregation. Formation of surfactant aggregates has been referred to briefly above. The driving force for aggregation in aqueous media is the extrusion of hydrocarbon chains from solution upon micelle formation, resulting in an overall decrease in the free energy of the system. In nonaqueous solution, aggregation of the surfactant molecules depends upon both the solvent and the surfactant structure.

Interactions between solvent and surfactant hydrocarbon chain groups tend to minimize the size of the aggregate, while interactions between the polar groups of the surfactants promote aggregation, in polar solvent.

Kinetic treatments in both aqueous and nonaqueous micellar systems have been based on the Hartley model (Hartley 1936, 1955) of opposing hydrophobic interactions and electrostatic repulsions which are responsible for micellization in water.

Surfactant association in apolar solvents is predominantly the consequence of dipole–dipole and ion pair interactions between the amphiphiles. This differs from the Hartley model and concepts derived for surfactant association in water may not necessarily be applicable to those in apolar solvents (O'Connor and Lomax 1983).

In a nonaqueous solution of concentration, C, existing as simple molecules, m, and micelles composed of n molecules in equivalent concentration, M_n, the mass law is as follows:

$$KM_n = m^n = (C - nM_n)^n \qquad (2.1)$$

where K is the dissociation constant of the micelles.

A phase separation model was advocated by Shinoda and Hutchinson (1962), and successfully applied by Singleterry (1955) and Fowkes (1962) to describe the aggregation of dinonylnaphthalene sulfonates in benzene. The phase separation model postulates that micellization is a phase transition. In its simplest form, it does not contain a size-limiting step and, therefore, it is of little value in accounting for the formation of the small aggregates seen in apolar media.

The mass action law can be applied to the overall aggregation process:

$$nm \overset{K_n}{\longleftrightarrow} M_n$$

where K_n is the association constant of the process that allows a model. Thus, with the conservation of mass:

$$[m]/[D]+n([m]/[D])^{n}[D]^{n-1}K_{n}=1 \qquad (2.2)$$

where $[D]$ is the total molal concentration of detergent. From this equation assuming that Kn is the product of $n-1$ individual and equal mass action constants, the following equation is obtained:

$$[m]/[D]+n[M_{n}]/[D]=1 \qquad (2.3)$$

and the CMC is then obtained:

$$CMC=1/k=[D] \qquad (2.4)$$

where $[D]=1$.

This system allows for two aggregation states: monomers and micelles. This does not account for the distribution in molecular weights. Smooth transitions from monomer, dimmer, trimer, etc., with concentration-dependent growth of aggregates have been observed. These gradual physical changes have been described (Lo et al. 1975) in terms of a sequential type self-association model. Assuming all values of equilibrium concentration are the same, K_{12}, K_{23}...K_{ij} are assumed to be equal:

$$Monomer + monomer \overset{K_{12}}{\leftrightarrow} dimer$$

$$Dimer + monomer \overset{K_{23}}{\leftrightarrow} trimer$$

$$(n-1)mer + monomer \overset{K_{ij}}{\leftrightarrow} n - mer$$

Then the weight fraction of the monomer, f, is related to the stoichiometric concentration of detergent, $[D]$, by the following equation:

$$(1-f^{1/2})/f = K_{ij}[D] \qquad (2.5)$$

It would be possible to modify the multiple equilibrium model to account for a critical concentration (Eike 1980). Application of the mass action law to aggregation, and conserving mass with respect to monomer yields

$$[m]/[D] + \sum_{n-2}^{n} n([m]/[D]^n)[D]^{n-1} \prod_{n-2}^{n} k_{n-1} = 1 \qquad (2.6)$$

A size-limiting step may be introduced by requiring a functional relationship between the equilibrium constants and the association number, $K_n = f(n)$ (Muto et al. 1974).

The number of monomers involved in most surfactant aggregates in nonpolar solvents is relatively small (typically less than 10 for alkylammonium carboxylates compared with up to 100 for aqueous micelles; Fendler and Fendler 1975); consequently, a spherical micelle structure would not provide effective shielding of the polar head groups from the solvent and its formation would be considered unlikely (Kertes and Gutmann 1976). The alternative model is that of a lamellar micelle in which the polar and hydrophobic groups are placed end to end and tail to tail, with water and organic solvents between them (Philippoff 1950; Mayer et al. 1969).

The kinetics of formation and decomposition of micelles and of the association–dissociation of the monomer to and from the micelle has rarely been studied in reversed micellar solutions (Yamashita et al. 1982). The paucity of data is attributed to: (1) the aggregation number of the micelle being very low that an abrupt change in the physico-chemical properties of the solution cannot be expected at the CMC and (2) micelle formation and monomer exchange reactions are too rapid to be observed by conventional techniques.

Adsorption from Solution

The method by which surfactants interact with other substances is adsorption at the interface. The interface may be a gas–liquid, liquid–liquid or solid–liquid juncture, the latter being significant in many pharmaceutical systems.

Langmuir presented a general equation for the isotherm of localized adsorption that was suitable for describing the adsorption of solutes. Langmuir's approach was concerned with monolayer coverage (Langmuir 1917). The assumptions made for this model were: that molecules are adsorbed at active centers on the adsorbent surface which they occupy for a finite period of time; owing to the small radius of action of adsorption forces and to their saturability, every active center while adsorbing molecules becomes incapable of further adsorption. Langmuir's adsorption equation concerns the simplifying assumptions that the heat of adsorption is independent of surface coverage, thus ignoring adsorbate interaction and the weakening effect on the intermolecular forces by distance between the adsorbent and adsorbate. Fowler and Guggenheim (1960) adopted an approach which provides a modification for lateral interactions in the Langmuir model. Attempts have been made to generalize monolayer and multilayer concepts in order to describe the isotherms of different shapes by a single equation. Brunauer et al. (1932) developed such a generalized theory in respect to adsorption of vapors which has since become known as the BET theory. The assumption of this theory are: the adsorbent surface has a definite number of active sites which are equivalent energetically and are capable of retaining the adsorbate molecules; the interaction of the neighboring adsorbed molecules is neglected; the molecules in each layer act as an adsorption site for subsequently adsorbed molecules; it is assumed that all of the adsorbed molecules in the second and subsequent layers have the same partition function as in the liquid state, which differs from the partition function of the first layer. Brunauer (1945) has classified adsorption isotherms into five types.

The BET theory has been very useful in the interpretation of solute adsorption. This theory assumes that every molecule of a liquid has only two close neighbors, from the top and bottom chain, while the molecules of a real liquid are surrounded by many more adjacent molecules. Moreover, Giles (Giles et al. 1974a,b) has found both theoretically and experimentally the shape of some isotherms, of solute from solution, can be accounted for by postulating that adsorbate interaction does occur under particular conditions. The assumption that the adsorbent surface has a definite number of active sites which are equivalent energetically is an oversimplification (Rudzinski and Narkiewicz-Michalek 1982, Mabire et al. 1984).

The Langmuir model for adsorption assumes that while the adsorbed molecules occupy sites of energy Q that they do not interact with each other. Fowler and Guggenheim (1960) have adopted an approach which accounts for lateral interaction.

The probability of a given site of energy Q being occupied is N/S, and if each site has z neighbors the probability of neighbor site being occupied is zN/S. So, the fraction of adsorbed molecules is $z\theta/2$, the factor one-half correcting for double counting and θ being total monolayer coverage. If the lateral interaction energy is ω, the added energy of adsorption is $z\omega\theta/2$, and the added differential energy of adsorption is just $z\omega\theta$.

The modified Langmuir equation then becomes

$$\Theta = b'C(1+b'C)^{-1} \tag{2.7}$$

$$b' = b_0\exp(Q+Zwq)/RT = b\,\exp(Zwq/RT) \tag{2.8}$$

rearranging these equations

$$bC = q/(1-q)\,\exp(-Zwq/RT) \tag{2.9}$$

this is equivalent to the Frumkin or Volmer (Damaskin et al. 1971) expression

$$bC = q/(1-q)\exp(-2aq) \tag{2.10}$$

therefore, $a=z\omega/2RT$, where a is the interaction energy between adsorbed molecules.

The adsorption of organic substances may yield sigmoid or logarithmic isotherms (Giles et al. 1974a), depending on the interaction between the adsorbed particles being predominantly attractive or repulsive.

The forces responsible for solute adsorption may be chemical or physico-chemical and physical or mechanical (Giles 1982). The chemical or physicochemical forces may be listed as covalent bonding; hydrogen bonds and other polar forces; ion exchange attraction; van der Waals forces; and hydrophobic forces. The physical or mechanical forces are: restriction of movement of solute aggregates in micropores and facilitation of entry of solute by the progressive breakdown of the substrate structure.

Solid Surface Interaction

Figure 2.1a illustrates the classical model of molecules associating at the surface of particles. However, it has long been known that surfaces exhibit roughness and asperities that contribute fundamentally to interactions. Figure 2.1b illustrates a more realistic depiction of the surface of a crystalline particle with areas of higher energy density (Hickey et al. 2007). The potential sources of these higher energy density sites may be the presence of amorphous material, moisture, impurities, electrical charge, and sites for mechanical interlocking. It should be evident from these images that depending on the scale of the means of measuring the surface, a realistic surface may seem bigger or smaller based on the ability to penetrate into the small invaginations in the surface. Driven by knowledge of the true nature of surfaces, morphologic approaches have included Fourier analysis and fractal analysis to

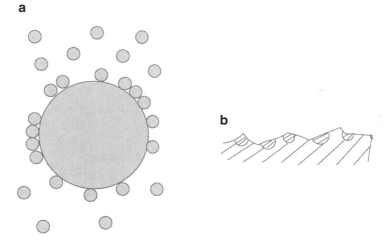

Fig. 2.1 (**a**) Classical model of surface adsorption phenomenon (*small circles* represent gas or solute molecules impinging at the surface of an idealized, spherical particle) and (**b**) schematic of particle surface depicting surface asperities (irregularity) and high energy density sites (indicated by *hatched hemispherical areas*)

approximate surfaces (Kaye 1993; Beddow 1976; Meloy 1977). Briefly, Fourier analysis involves describing a particle surface by assigning a center to the image of the particle and then using polar coordinates to plot variations of the surface on a linear scale (Luerkens 1991). This image can then be subjected to harmonic analysis from which a Fourier series and the respective coordinates each corresponding with a shape, sphericity, triangularity, etc., can be derived to describe the particle based on its surface. The principle of fractal geometry is that when periphery of an object, such as a particle, is measured at different scales of scrutiny, with yardsticks of different length, it appears to be larger as the scale is smaller (Kaye 1989). There is a linear relationship between the exponent of the estimate of periphery and that of the dimension of the scale being employed. From the slope of this line the so-called fractal dimension can be derived. This is often thought of as representing dimensions between 1 and 2, for flat, 2D images and between 2 and 3 for 3D images.

Beyond the morphology of individual particles powders, composed of numerous particles, have been probed with molecules, each with a radius that establishes the scale of scrutiny to derive fractal dimensions (Avnir et al. 1984; Pfeifer and Avnir 1983; Avnir and Farin 1983; Avnir 1989). These experiments which employ inert gas molecules of different dimensions have been extended to consider the functional status of the surface. It could be argued that the use of the Scatchard interpretation of a Langmuirian adsorption but based on the specific binding capacity of proteins was an early example of a functional surface interaction (Xu et al. 2010). However, techniques such as inverse phase gas chromatography (Telko and Hickey 2007) and atomic force microscopy (Danesh et al. 2000) allow direct probing of surface features with respect to the force or energy associated with potential molecular interaction.

The recent extension of this molecule–particle interaction to particle–particle interactions may lead to greater understanding of the complex interactions of particles in pharmaceutical formulation which will be take up in great detail in Sect. 3.

Summary

Chemical reactions are typified by apparently predictable behavior which masks a complex range of underpinning states. Interest in this subject is beginning to emerge in chemistry and may ultimately be useful in considering subtleties of chemical stability of pharmaceuticals (Carstensen 1990). The nature of surfaces and their potential to interact with molecules or particles stems from a long and extensive history of attempts to model nonlinear phenomena in surface and interfacial chemistry. The strong foundation in this field gives a major opportunity for new developments as complex interpretations are employed to elucidate surface phenomena. The implications of greater understanding in this area with respect to important pharmaceutical phenomena such as deaggregation, dissolution, diffusion, and ultimately drug availability are taken up in the next section.

References

Avnir D (1989) The fractal approach to heterogeneous chemistry, surfaces, colloids, polymers. Wiley, New York

Avnir D, Farin D (1983) Chemistry in noninteger dimensions between two and three. II. Fractal surfaces of adsorbents. J Phys Chem 79:3566–3571

Avnir D, Farin D, Pfiefer P (1984) Molecular fractal surfaces. Nature 308:261–263

Beddow JK, Vetter AF, Sisson K (1976) Powder metallurgy review 9, Part I, Particle shape analysis. Powder Metall Int 8:69–76

Billing GD, Mikkelsen KV (1996) Introduction to molecular dynamics and chemical kinetics. Wiley, New York

Bonchev D, Kamenski D, Temkin ON (1987) Complexity index for linear mechanisms of chemical reactions. J Math Chem 1:345–388

Brunauer S (1945) The adsorption of gases and vapors, I. Princeton University Press, Princeton

Brunauer S, Emmett PH, Teller E (1932) Adsorption of gases in multimolecular layers. J Am Chem Soc 6:309–319

Carstensen JT (1990) Drug stability principles and practices. Dekker, New York

Damaskin BB, Petril OA, Batrakov VV (1971) Adsorption of organic compounds on electrodes. Plenum, New York

Danesh A, Chen X, Davies MC, Roberts CJ, Sanders GHW, Tendler SJB, Williams PM, Wilkins MJ (2000) The discrimination of drug polymorphic forms from single crystals using atomic force microscopy. Pharm Res 17:887–890

De Oliveira CR, Werlang T (2007) Ergodic hypothesis in classical statistical mechanics. Rev Bras Ensino Fis 29:189–201

Eike HF (1977) Micelles in apolar media. In: Mittal KL (ed) Micellization, solubilization and microemulsions. Plenum, New York

Eike HF (1980) Aggregation in surfactant solutions: formation and properties of micelles and microemulsions. Pure Appl Chem 52:1349–1357

Fendler JH, Fendler EJ (1975) Catalysis in micellar and macromolecular systems. Academic, New York

Fowkes FM (1962) The micelle phase of calcium dinonylnaphthalene sulfonate in n-decane. J Phys Chem 66:1843–1845

Fowler R, Guggenheim EA (1960) Statistical thermodynamics. Cambridge University Press, Cambridge

Giles CH (1982) Forces operating in adsorption of surfactants and other solutes at solid surfaces: a survey. In: Mittal KL, Fendler EJ (eds) Solution behavior of surfactants theoretical and applied aspects. Plenum, New York

Giles CH, D'Silva AP, Easton IA (1974a) A general treatment and classification of the solute adsorption isotherm. I. Theoretical. J Colloid Interface Sci 47:755–765

Giles CH, D'Silva AP, Easton IA (1974b) A general treatment and classification of the solute adsorption isotherm. II. Experimental interpretation. J Colloid Interface Sci 47:766–778

Hartley GS (1936) Aqueous solutions of paraffin chain salts. Hermann and Cie, Paris

Hartley GS (1955) Progress in chemistry of fats and other lipids. Pergamon, London

Hickey AJ, Mansour HM, Telko MJ, Xu Z, Smyth HD, Mulder T, McLean R, Langridge J, Papadopoulos D (2007) Physical characterization of component particles included in dry powder inhalers. II. Dynamic characteristics. J Pharm Sci 96:1302–1319. doi:10.1002/jps.20943

Kaye BH (1989) A random walk through fractal dimensions. VCH Publishers, New York

Kaye BH (1993) Applied fractal geometry and the fine particle specialist. Part I. Rugged boundaries and rough surfaces. Part Part Syst Charact 10:99–110

Kertes AS (1977) Aggregation of surfactants in hydrocarbons, incompatibility of the critical micelle concentration concept with experimental data. In: Mittal KL (ed) Micellization, solubilization and microemulsions. Plenum, New York

Kertes AS, Gutmann H (1976) Surfactants in organic solvents: the physical chemistry of aggregation and micellization. In: Matijevic E (ed) Surface and colloid science, vol 8. Wiley, New York

Langmuir I (1917) The constitution and fundamental properties of solids and liquids. II. Liquids. J Am Chem Soc 39:1848–1906

Lo FYF, Escott BM, Fendler EJ, Adams ET, Larson RD, Smith PW (1975) Temperature-dependent self-association of dodecylammonium propionate in benzene cyclohexane. J Phys Chem 79:2609–2621

Luerkens DW (1991) Theory and applications of morphological analysis, fine particles and surfaces. CRC Press, Boca Raton

Mabire F, Audebert R, Quivoron C (1984) Flocculation properties of some water soluble cationic copolymers towards silica suspensions: a semiquantitative interpretation of the role of molecular weight and cationicity through a "patchwork" model. J Colloid Interface Sci 97:122–136

Mayer I, Gutmann H, Kertes AS (1969) In: Kertes AS, Marcus Y (eds) Solvent extraction research. Wiley-Interscience, New York

Meloy TP (1977) Fast Fourier transforms applied to shape analysis of particle silhouettes to obtain morphological data. Powder Technol 17:27–35

Mukerjee P (1974) Micellar properties of drugs: micellar and non-micellar patterns of self association of hydrophobic solutes of different molecular structures, monomer fraction, availability, and misuses of micellar hypotheses. J Pharm Sci 63:972–981

Muto S, Shimasaki Y, Meguro K (1974) The effect of counterion on critical micelle concentration of surfactant in non-aqueous media. J Colloid Interface Sci 49:173–176

Nowak G, Fic G (2010) Search for complexity generating chemical transformations by combining connectivity analysis and cascade transformation patterns. J Chem Inf Model 50:1369–1377. doi:10.1021/ci100146n

O'Connor CJ, Lomax TD (1983) Evidence for the sequential self-association model in reversed micelles. Tetrahedron Lett 24:2917–2920

Pfeifer P, Avnir D (1983) Chemistry in noninteger dimensions between two and three. I. Fractal theory of heterogeneous surfaces. J Chem Phys 79:3558–3565

Philippoff W (1950) Micelles and X-rays. J Colloid Sci 5:169–191

Ravey JC, Buzier M, Picort C (1984) Micellar structures of non-ionic surfactants in apolar media. J Colloid Interface Sci 97:9–25

Rudzinski W, Narkiewicz-Michalek J (1982) Adsorption from solutions onto solid surfaces. J Chem Soc Faraday Trans I 78:2361–2368

Shinoda K, Hutchinson E (1962) Pseudo-phase separation model for thermodynamic calculations on micellar solutions. J Phys Chem 66:577–582

Singleterry CR (1955) Micelle formation and solubilization in non-aqueous solvents. J Am Oil Chem Soc 32:446–452

Szasz D (1994) Boltzmann's ergodic hypothesis, a conjecture for centuries? Preprint ESI(8). The Erwin Schrödinger International Institute for Mathematical Physics, Wien, Austria

Tanford C (1980) The hydrophobic effect, 2nd edn. Wiley, New York

Telko MJ, Hickey AJ (2007) Critical assessment of inverse gas chromatogra-
    phy as means of assessing surface free energy and acid–base interactions
    of pharmaceutical powders. J Pharm Sci 96:2647–2654
Xu Z, Mansour HM, Mulder T, McLean R, Langridge J, Hickey AJ (2010)
    Heterogeneous particle deaggregation and its implication for therapeutic
    aerosol performance. J Pharm Sci 99:3442–3461. doi:10.1002/jps.22057
Yamashita T, Yano H, Harada S, Yasunaga T (1982) Kinetic studies of the
    micelle system of octylammonium alkanoates in hexane solution by the
    ultrasonic absorption. Bull Chem Soc Jpn 55:3403–3406

Chapter 3
Solid State Pharmaceuticals: Solving Complex Problems in Preformulation and Formulation

The preformulation and formulation of new chemical entities (NCEs) have evolved considerably in the past 50 years, being driven largely by dramatic changes in the regulation of drug products. The objective of this chapter is to discuss examples of several well-known and ubiquitous issues in preformulation/formulation data analysis and interpretation. These issues have proved extremely challenging for scientists as they have attempted to develop drugs and drug products due to the inability of standard modeling or traditional mathematical approaches to mimic the processes or provide a predictive way forward. However, interpretation of the problem using different approaches originating from the fields of dynamical systems, chaos, complexity and fractals, tenable solutions, and even deeper understanding has been facilitated. The chapter is divided into two sections: "Statics" and "Dynamics." Statics relates to the static properties of NCEs that are generally considered to be physical and chemical properties that are determined during preformulation. When the NCEs enter formulation stages of development, one has to consider more dynamic relationships of the drug and delivery system. This important distinction between static and dynamic systems with respect to complexity and data interpretation will allow readers to not only consider these presented issues but also extend these approaches to analysis to their own specific data set.

A.J. Hickey and H.D.C. Smyth, *Pharmaco-Complexity*, Outlines in Pharmaceutical Sciences 1, DOI 10.1007/978-1-4419-7856-1_3, © American Association of Pharmaceutical Scientists 2011

Statics

Following lead optimization, preformulation studies are critical for candidate selection. This thorough characterization of the physicochemical properties of drugs is the foundation for developing robust dosage forms and ensuring development times and costs are optimized. Most developers of new drugs will perform preformulation studies that address several key physicochemical properties that are known to have a significant impact on the manufacturing process and the dynamics of the formulation. Characteristics of drugs determined during preformulation studies will give some indication of the likelihood of "druggability" or formulation success but not in all cases (e.g., Lipinski's rule of five) (Lipinski et al. 2001; Keller et al. 2006) The focus of this subsection is to highlight some aspects of the physicochemical properties of drugs in which clear known examples of complexity have been established, and discuss briefly the approaches taken by scientists to cope with this behavior.

Solubility, Dissolution, and Release

Prediction of solubility, dissolution, and drug release is of great interest in the pharmaceutical industry. Solubility, for example, is now widely recognized as an important issue preventing drug development, largely because compounds are often selected from databases and libraries, and judged oftentimes purely from their structures (Delaney 2004, 2005). Drug dissolution is key for drug absorption and therefore clinical effect. Thus, it is a paramount concern that needs to be addressed early in pharmaceutical product development. In 1897, Noyes and Whitney noticed that the rate of dissolution is proportional to the difference between the instantaneous concentration, C at time t, and the saturation solubility, C_S (Dokoumetzidis and Macheras 2006).

$$\frac{dC}{dt} = k(C_s - C) \tag{3.1}$$

where k is a constant. In the pharmaceutical field, once dissolution was established to be an important issue for drug absorption and bioavailability, many aspects of this process were investigated. Solubility, stirring, particle size, and wettability were among the key aspects studied by the rapidly expanding field of dissolution science. Initially, these studies focused on the changes that could be induced in dissolution processes by modifying the environment and formulations. Subsequent emphasis has been on dissolution as a predictor of oral drug absorption and this led to a landmark paper introducing the Biopharmaceutics Classification System (BCS) by Amidon et al. (1995). Many studies then were published introducing new models that attempted to correlate in vitro studies to in vivo observations (e.g., Dressman et al. 1998; Persson et al. 2005). Despite advances in dissolution testing provided by these and other studies, the field has yet to overcome obstacles to modeling the in vivo situation. Specific obstacles in this field are the complexity and chaotic nature of the hydrodynamic conditions of dissolution in vivo (D'Arcy et al. 2005, 2006), the complexity of gastrointestinal drug absorption phenomena (Macheras and Argyrakis 1997), and heterogeneity of in vivo conditions. An excellent review of the history of dissolution research was recently provided by Dokoumetzidis and Macheras (2006). Establishing models and tests for predicting dissolution through different methods is clearly needed.

Fractal Dimensions and Surface Phenomena

An introduction to fractals and fractal geometry is provided elsewhere in this text and in excellent texts on the subject (Barnsley and Rising 1993; Kaye 1994). Examples are shown in Fig. 3.1 and basic definitions given in Table 3.1. The application of fractals to surfaces has been largely facilitated by the work of Avnir and colleagues. They explored fractal geometry in chemistry and physics, applying this mathematical approach to surface interactions (Avnir et al. 1984). Avnir and Farin showed that at the molecular level, the surfaces of most materials were fractal. They extended their work in fractal chemistry to drug dissolution. Notably, they provided a modified Noyes–Whitney equation and a Hixson–Crowell cube root

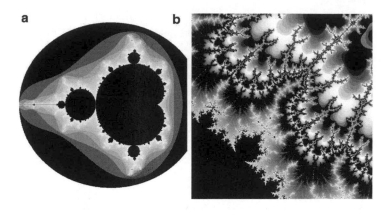

Fig. 3.1 Example of a fractal (Mandlebrot set) viewed at a distance (**a**) and at much higher magnification (**b**) indicating the appearance of self-similarity at these different scales (images produced on FractalWorks software)

law to include surface roughness effects on the dissolution rate of drugs (Farin and Avnir 1992). Carstensen and Franchini (1993), however, argue that self-similarity of particles during dissolution may not be a valid assumption. However, several authors have shown the applicability of fractals to dissolution of pharmaceutical systems, for example, diclofenac (Holgado et al. 1995), sodium cholate (Fini et al. 1996), orthoboric acid (Tromelin et al. 1996), and others (Akbarieh et al. 1987; Akbarieh and Tawashi 1989; Farin and Avnir 1992; Schroder and Kleinebudde 1995; Weidler et al. 1998; Johns and Gladden 2000). Recently, these models have been extended to better predict solubility of nanoparticles compared to the Ostwald–Freundlich equation (Mihranyan and Strømme 2007).

Monte Carlo Methods in Modeling Drug Release

Since the 1970s, sustained release delivery systems have evolved rapidly. The release of drug from these systems must be predictable for the delivery system to be successful. The kinetics of drug release follows the operative release mechanism of the system, for example, diffusion through inert matrix, diffusion across membrane or hydrophilic gel, osmosis, and ion exchange. By far, diffusion is

Table 3.1 Fractal geometry primer

Definition	A fractal is an object/quantity that displays self-similarity across many scales of scrutiny. The object need not exhibit exactly the same structure at all scales, but the same "type" of structures must appear on all scales. A plot of the quantity on a log–log graph versus scale gives a straight line, whose slope is said to be the fractal dimension (Weisstein 2010).	
Example	The length of a coastline measured with different length rulers. The shorter the ruler, the longer the length measured, a paradox known as the coastline paradox.	
Examples in pharmaceutical sciences	Dissolution (Akbarieh and Tawashi 1989; Farin and Avnir 1992; Tromelin et al. 1996, 2001)	Particle analysis (Avnir et al. 1991; Concessio and Hickey 1997; Fini et al. 2002)
	Imaging (Zook and Iftekharuddin 2005; Martin-Landrove et al. 2007)	Drug distribution, pharmacokinetics, and toxicity (Macheras 1996; Fuite et al. 2002; Karalis and Macheras 2002; Marsh and Tuszynski 2006; Marsh and Riauka 2007; Pang et al. 2007; Pereira 2010)
	Drug diffusion (Roncaglia et al. 1994; Liu and Nie 2001)	Pharmacodynamic responses (Jartti et al. 1998)

the principal release mechanism. Solute release models were described early and modeling drug release from diffusion-controlled systems relies on the Higuchi model published in 1961 (Higuchi 1961). The kinetics of release from an ointment was described, assuming homogeneously mixed drug in a planar matrix diffusing into a medium, a perfect sink, under pseudo steady-state conditions. This model has been widely used and was instrumental for the development and understanding of dosage form design. However, due to the approximate nature of the model, its use for the analysis of release data is recommended only for the first 60% of the release curve, beyond which, the model is insufficient. In 1985, Peppas introduced a semiempirical equation (the power law) to describe drug release from polymeric devices (Peppas 1985; Siepmann and Peppas 2001). Similarly, valid estimates of drug release can be obtained by fitting this equation to the first 60% of the experimental release data. This highly cited work has been widely applied in the analysis of drug release studies. Mechanistic release models have been published in literature (Siepmann and Peppas 2001) and are more physically realistic, but their mathematical complexity is their main disadvantage for wide use.

More recently, the methodology used to find the release mechanisms for an entire set of data has been demonstrated via the use of Monte Carlo simulations. Kosmidis and coworkers (Kosmidis et al. 2003b; Kosmidis and Macheras 2007, 2008) showed that the Weibull function could be used to describe release kinetics in either Euclidean or fractal spaces. Based on these findings, a new methodology was developed for the description and determination of release mechanisms using the *entire* set of data (Papadopoulou et al. 2006) (Table 3.2).

Surfaces and Particles

Complexity of Crystallization

Over 90% of all pharmaceutical products contain active ingredients produced in crystalline form. Crystallization processes can illustrate some interesting dynamical behavior, including a high sensitivity to parameter variations. This can be a cause of considerable issues in

Table 3.2 Monte Carlo method

Definition	The use of randomly generated or sampled data and computer simulations to obtain approximate solutions to complex mathematical and statistical problems. Monte Carlo methods enable the investigation of complex and dynamical systems by direct study of their properties using computer simulation.	
Examples in pharmaceutical sciences	Drug release (Kosmidis et al. 2003a; Kosmidis and Macheras 2008)	Colloidal dispersions (Sjoberg and Mortensen 1997; Aoshima and Satoh 2005; Narambuena et al. 2005; Wang et al. 2009; Sanz and Marenduzzo 2010)	Patient compliance (Ahmad et al. 2005)
	Solubility prediction (Jorgensen and Duffy 2000)	Pharmacokinetics (Montgomery et al. 2001; Dokoumetzidis et al. 2005)	Manufacturing and processing (Kuu and Chilamkurti 2003; Rowe et al. 2005)

production and also a significant impact on the efficiency and profitability of the overall crystallization process. Differences and variability in the production of the crystals can impact downstream processes too (often in a nonlinear and complex way): filtration, drying, milling, powder flow, surface energetics, bioavailability, tablet stability, etc. The dynamics of both continuous crystallization and batch crystallization processes have been investigated and reviewed (Rawlings et al. 1993; Braatz and Hasebe 2002). Crystallization processes are highly nonlinear and are modeled by coupled nonlinear algebraic integro-partial differential equations. However, due to large differences and length scales (Angstrom to micron) and time scales (microseconds to minutes), these equations are found not to be useful for nonlinear feedback controllers using available computers (Christofides 2002). Therefore, the use of tools from nonlinear dynamics and complexity has been described for this problem and summarized in a recent review by Fujiwara and coworkers (Fujiwara et al. 2005).

Complexity of Particle Shape

Particle morphology and particle size distributions can have profound effects on the manufacture, processing, and performance of pharmaceutical dosage forms. The shape of particles within a powder is rarely regular, but it is very valuable to be able to describe particle shape using quantitative measures. This can enable understanding of dynamics of drug dissolution, aerodynamics, and other important properties conferred by shape in pharmaceutical systems. It is also necessary to have a quantitative assessment of shape for quality control and regulatory purposes. As such, a variety of shape factors have been described in the literature (Concessio and Hickey 1997). Static shape factors are those that can be obtained from imaging of particles and they attempt to describe either the deviation from ideal shape (e.g., sphere) or shape-independent of a physical reference. The most useful and widely described factor in the modern literature is fractal analysis of particle shape. Fractal geometry and fractal surfaces are based on self-similarity of a surface, or features, at different scales. As one looks at an object at increasing or decreasing levels of scrutiny, the features and patterns are observed to be repetitive. Rugged structures such as powder particles can be assessed using fractal geometry via estimation of a mass fractal dimension or boundary fractal dimension (Kaye 1978). Different methods have attempted to characterize particle shape using polar coordinates from which Fourier coefficients are derived (Luerkens 1991). However, these methods are tedious and use of surface area determinations across narrow particle size distributions has been suggested to be an alternative method of obtaining a fractal dimension estimate of particle shape (Concessio and Hickey 1997).

Dynamics

This section will focus on solids and powders, their processing and formulation.

Powder Flow and Mixing

The dynamic behavior of powders and granular matter is a well-known example of collective systems that are far from equilibrium (Ottino and Khakhar 2002; Muzzio et al. 2003). They have been used to illustrate nonlinear dynamics and complex systems as systems that experience self-organization, invariance and symmetry breaking, and pattern formation (waves, chaos) (Hill et al. 1999; Khakhar et al. 1999; Ottino and Khakhar 2002; Gilchrist and Ottino 2003). Moreover, these concepts have also been applied across a wide range of scales from pharmaceutical fine particles to the large scale geological movement of ice floes.

Clearly, the flow properties of powders are of great significance for the transfer, sampling, and mixing of pharmaceutical materials. In addition, researchers are interested in using powder flow to probe interactions between particles within a powder. The underlying physical phenomenon of interparticulate interactions have been correlated to Carr's compressibility index (Carr 1965),

Hausner's ratio (Hausner 1967), and other methods of powder flow: Kawakita's constant (Lüdde and Kawakita 1966), shear cell measurement (Carr and Walker 1967), critical orifice diameter (Walker 1966), and rotating drums (Concessio et al. 1999; Lee et al. 2000; Castellanos et al. 2002; Crowder and Hickey 2006; Faqih et al. 2006). Static measures of powder flow give measurable parameters that describe a powder's ability to flow under certain conditions. It is possible to predict the flow behavior of the powder but the prediction depends on the relationship between each method of quantification and the process being predicted. For this reason, dynamic methods of powder flow have been developed such as flow during vibration, dynamic flow using texture analyzers and strain gages, and rotating drum methods.

During powder flow, the powder must be expanded and interparticulate forces must be overcome. Particles then increase their separation distance allowing the static system to become more like a fluid. The flow of powders may be characterized in four phases – plastic solid, inertial, fluidization, and suspension (Crowder et al. 2003) and these phases correlate to the particle spacing, interparticulate forces, and degree of mobility of individual particles in the system. Powders are composed of millions of particles, each of different morphology, size, and other physical properties, and observing flow can be interpreted as random. However, these systems have been suggested to have underlying order during flow (Hickey and Concessio 1996). Apparent disorder in a chaotic system can be due to a large number of unstable periodic motions (Grebogi et al. 1988). In early studies on pharmaceutically relevant powders, Hickey and Concessio showed that using a rotating drum underlying order in powder flow could be detected. As a powder sample is slowly rotated in a drum, the powder rises until its angle of repose is exceeded and an avalanche occurs (Hickey and Concessio 1996). One can measure the time between avalanching or the dynamic angle of repose. The change of angle of repose with respect to the mean angle plotted against time produces an oscillating data plot. A phase space attractor plot (Concessio and Hickey 1997) can be generated where data points cluster around a central attractor point. The scatter around this attractor point represents a measure of variability in the avalanching behavior. Lower attractor

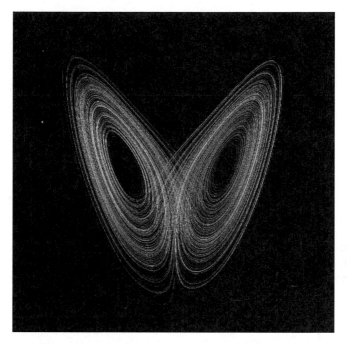

Fig. 3.2 Lorenz attractor, an example of an attracting set, that has zero measure in the embedding phase space and has fractal dimension and evolves over time (Anonymous 2010)

points indicate better flowability (Kaye 1993). These oscillating data and phase space attractor plots can be used as indicators of deterministic chaos, through which patterns different powder behaviors can be distinguished (Fig. 3.2).

Related to the process of flow is the ubiquitous process of mixing and blending in the pharmaceutical sciences. The dynamics of mixing and de-mixing is well known to pharmaceutical scientists and is considered a problem in most cases as it is often hard to predict when a multicomponent mixture will become homogenous and when it may change into a segregated state. Clearly, this has implications for the overarching theme of uniformity of the dosage form that is required of pharmaceutical systems (Venables and Wells 2001; Shah et al. 2007). In many cases, de-mixing occurs due to size segregation. Particles may reorganize via the gaps found around larger particles, through which smaller particles may slip during energy input into the powder (vibrations, movement, etc.). The lower density of smaller particles at the top of a powder bed then allows the large particles to move upward. In addition, segregation has been well characterized for systems where components have different densities and can be seen in pharmaceutical systems and is now being investigated from a theoretical perspective during manufacturing and processing (Egermann et al. 1992; Xie et al. 2008). The general approach to ensure uniform mixing or avoidance of segregation in pharmaceutical processing has been through judicious selection of excipients and processing conditions without development of models or assessment of the dynamics of the system. Therefore, recent developments in the mathematics of nonlinear dynamical systems used to describe spontaneous pattern forming, "emergence," and self assembly have enabled improved understanding and potentially control over mixing. Spontaneous chaotic mixing of powders less than 300 μm diameter was first reported by Shinbrot et al. (1999). Prior to this report, it was thought that mixing in granular flows was thought to be diffusive. Periodic slipping and sticking of the powders blended in cylindrical tumblers are the mechanisms by which this chaotic mixing occurs and allows much more rapid mixing rates to occur (Table 3.3).

Table 3.3 Other applications of pharmacocomplexity in formulation development

Problem	References
Particle morphology and dissolution	Farin and Avnir (1992), Concessio and Hickey (1997), Dokoumetzidis and Macheras (2006)
Emergence of order in oscillated powders	Moon et al. (2004)
Fluidized beds	Daw et al. (1995)
Milling	Manai et al. (2002)
Controlled drug delivery	Li and Siegel (2000)
Mixing	Shinbrot et al. (1999), Ottino and Khakhar (2002), Christov et al. (2010)
Tablet compaction	Leuenberger et al. (1992)

Summary

Drug discovery efforts lead to an ever-expanding array of physicochemically diverse compounds that need to be characterized and formulated into an appropriate dosage form. Historically well-described issues in formulation development (e.g., solubility, dissolution, particle size analysis, and mixing) and emerging issues (manufacturing, scale-up, excipient interactions, modulated and controlled release systems, etc.) are often characterized by their nonlinearity and complexity. Despite frequently originating from seemly simple systems and regardless of our apparent increases in data/knowledge collection, the systems have remained unpredictable in many cases. There are many examples already in preclinical development of the usefulness of the tools of pharmacocomplexity. It has probably been due to the intersection of the field of physical pharmacy with fields of physics, engineering, and chemistry that has seen the introduction of many of the concepts of complexity, chaos, and fractals and that this is anticipated to continue, perhaps even through more formalized training of those entering the discipline.

References

Ahmad AM, Douglas Boudinot F et al (2005) The use of Monte Carlo simulations to study the effect of poor compliance on the steady state concentrations of valproic acid following administration of enteric-coated and extended release divalproex sodium formulations. Biopharm Drug Dispos 26(9):417–425

Akbarieh M, Tawashi R (1989) Surface studies of calcium oxalate dihydrate single crystals during dissolution in the presence of stone-formers' urine. Scanning Microsc 3(1):139–145, discussion 145–136

Akbarieh M, Dubuc B et al (1987) Surface studies of calcium oxalate dihydrate single crystals during dissolution in the presence of urine. Scanning Microsc 1(3):1397–1403

Amidon GL, Lennernas H et al (1995) A theoretical basis for a biopharmaceutic drug classification: The correlation of in vitro drug product dissolution and in vivo bioavailability. Pharm Res 12(3):413–420

Anonymous (2010) http://commons.wikimedia.org/wiki/File:Lorenz_system_r28_s10_b2-6666.png. Accessed 22 Aug 2010

Aoshima M, Satoh A (2005) Two-dimensional Monte Carlo simulations of a colloidal dispersion composed of polydisperse ferromagnetic particles in an applied magnetic field. J Colloid Interface Sci 288(2):475–488

Avnir D, Farin D et al (1984) Molecular fractal surfaces. Nature 308(5956):261–263

Avnir D, Carberry JJ et al (1991) Fractal analysis of size effects and surface morphology effects in catalysis and electrocatalysis. Chaos 1(4):397–410

Barnsley MF, Rising H (1993) Fractals everywhere. Academic Press Professional, Boston

Braatz RD, Hasebe S (2002) Particle size and shape control in crystallization processes. In: AIChE symposium, series: proceedings of the 6th international conference on chemical process control

Carr RL (1965) Evaluating flow properties of solids. Chem Eng 72:163–168

Carr JF, Walker DM (1967) An annular shear cell for granular materials. Powder Technol 68(1):369–373

Carstensen JT, Franchini M (1993) The use of fractal geometry in pharmaceutical systems. Drug Dev Ind Pharm 19(1–2):85–100

Castellanos A, Valverde JM et al (2002) Fine cohesive powders in rotating drums: transition from rigid-plastic flow to gas-fluidized regime. Phys Rev E Stat Nonlin Soft Matter Phys 65(6 pt 1):061301

Christofides PD (2002) Model-based control of particulate processes. Kluwer Academic, The Netherlands

Christov IC, Ottino JM et al (2010) Chaotic mixing via streamline jumping in quasi-two-dimensional tumbled granular flows. Chaos 20(2):023102

Concessio NM, Hickey AJ (1997) Descriptors of irregular particle morphology and powder properties. Adv Drug Deliv Rev 26(1):29–40

Concessio NM, VanOort MM et al (1999) Pharmaceutical dry powder aerosols: correlation of powder properties with dose delivery and implications for pharmacodynamic effect. Pharm Res 16(6):828–834

Crowder T, Hickey A (2006) Powder specific active dispersion for generation of pharmaceutical aerosols. Int J Pharm 327(1–2):65–72

Crowder T, Hickey A et al (2003) A guide to pharmaceutical particulate science. Informa Healthcare, New York

D'Arcy DM, Corrigan OI et al (2005) Hydrodynamic simulation (computational fluid dynamics) of asymmetrically positioned tablets in the paddle dissolution apparatus: impact on dissolution rate and variability. J Pharm Pharmacol 57(10):1243–1250

D'Arcy DM, Corrigan OI et al (2006) Evaluation of hydrodynamics in the basket dissolution apparatus using computational fluid dynamics – dissolution rate implications. Eur J Pharm Sci 27(2–3):259–267

Daw CS, Finney CEA et al (1995) Self-organization and chaos in a fluidized bed. Phys Rev Lett 75(12):2308

Delaney JS (2004) ESOL: estimating aqueous solubility directly from molecular structure. J Chem Inf Comput Sci 44(3):1000–1005

Delaney JS (2005) Predicting aqueous solubility from structure. Drug Discov Today 10(4):289–295

Dokoumetzidis A, Macheras P (2006) A century of dissolution research: from Noyes and Whitney to the biopharmaceutics classification system. Int J Pharm 321(1–2):1–11

Dokoumetzidis A, Kosmidis K et al (2005) Modeling and Monte Carlo simulations in oral drug absorption. Basic Clin Pharmacol Toxicol 96(3):200–205

Dressman JB, Amidon GL et al (1998) Dissolution testing as a prognostic tool for oral drug absorption: immediate release dosage forms. Pharm Res 15(1):11–22

Egermann H, Krumphuber A et al (1992) Novel approach to estimate quality of binary random powder mixtures: samples of constant volume. III: Range of validity of equation. J Pharm Sci 81(8):773–776

Faqih A, Chaudhuri B et al (2006) An experimental/computational approach for examining unconfined cohesive powder flow. Int J Pharm 324(2):116–127

Farin D, Avnir D (1992) Use of fractal geometry to determine effects of surface morphology on drug dissolution. J Pharm Sci 81(1):54–57

Fini A, Fazio G et al (1996) Fractal analysis of sodium cholate particles. J Pharm Sci 85(9):971–975

Fini A, Holgado MA et al (2002) Ultrasound-compacted indomethacin/polyvinylpyrrolidone systems: effect of compaction process on particle morphology and dissolution behavior. J Pharm Sci 91(8):1880–1890

Fuite J, Marsh R et al (2002) Fractal pharmacokinetics of the drug mibefradil in the liver. Phys Rev E Stat Nonlin Soft Matter Phys 66(2 pt 1):021904

Fujiwara M, Nagy ZK et al (2005) First-principles and direct design approaches for the control of pharmaceutical crystallization. J Process Control 15(5):493–504

Gilchrist JF, Ottino JM (2003) Competition between chaos and order: mixing and segregation in a spherical tumbler. Phys Rev E Stat Nonlin Soft Matter Phys 68(6 pt 1):061303

Grebogi C, Ott E et al (1988) Unstable periodic orbits and the dimensions of multifractal chaotic attractors. Phys Rev A 37(5):1711

Hausner HH (1967) Friction conditions in a mass of metal powder. Int J Powder Metall 3:7–13

Hickey A, Concessio NM (1996) Chaos in rotating lactose beds. Part Sci Technol 14(1):15–25

Higuchi T (1961) Rate of release of medicaments from ointment bases containing drugs in suspension. J Pharm Sci 50:874–875

Hill KM, Khakhar DV et al (1999) Segregation-driven organization in chaotic granular flows. Proc Natl Acad Sci U S A 96(21):11701–11706

Holgado MA, Fernández-Hervás MJ et al (1995) Characterization study of a diclofenac salt by means of SEM and fractal analysis. Int J Pharm 120(2):157–167

Jartti TT, Kuusela TA et al (1998) The dose–response effects of terbutaline on the variability, approximate entropy and fractal dimension of heart rate and blood pressure. Br J Clin Pharmacol 45(3):277–285

Johns ML, Gladden LF (2000) Probing ganglia dissolution and mobilization in a water-saturated porous medium using MRI. J Colloid Interface Sci 225(1):119–127

Jorgensen WL, Duffy EM (2000) Prediction of drug solubility from Monte Carlo simulations. Bioorg Med Chem Lett 10(11):1155–1158

Karalis V, Macheras P (2002) Drug disposition viewed in terms of the fractal volume of distribution. Pharm Res 19(5):696–703

Kaye BH (1978) Specification of the ruggedness and/or texture of a fine particle profile by its fractal dimension. Powder Technol 21(1):1–16

Kaye BH (1993) Chaos & complexity: discovering the surprising patterns of science and technology. VCH, Weinheim; New York

Kaye BH (1994) A random walk through fractal dimensions. VCH, Weinheim; New York

Keller TH, Pichota A et al (2006) A practical view of 'druggability'. Curr Opin Chem Biol 10(4):357–361

Khakhar DV, McCarthy JJ et al (1999) Chaotic mixing of granular materials in two-dimensional tumbling mixers. Chaos 9(1):195–205

Kosmidis K, Macheras P (2007) Monte Carlo simulations for the study of drug release from matrices with high and low diffusivity areas. Int J Pharm 343(1–2):166–172

Kosmidis K, Macheras P (2008) Monte Carlo simulations of drug release from matrices with periodic layers of high and low diffusivity. Int J Pharm 354(1–2):111–116

Kosmidis K, Argyrakis P et al (2003a) A reappraisal of drug release laws using Monte Carlo simulations: the prevalence of the Weibull function. Pharm Res 20(7):988–995

Kosmidis K, Rinaki E et al (2003b) Analysis of Case II drug transport with radial and axial release from cylinders. Int J Pharm 254(2):183–188

Kuu WY, Chilamkurti R (2003) Determination of in-process limits during parenteral solution manufacturing using Monte Carlo Simulation. PDA J Pharm Sci Technol 57(4):263–276

Lee YS, Poynter R et al (2000) Development of a dual approach to assess powder flow from avalanching behavior. AAPS PharmSciTech 1(3):E21

Leuenberger H, Leu R et al (1992) Application of percolation theory and fractal geometry to tablet compaction. Drug Dev Ind Pharm 18(6–7):723–766

Li B, Siegel RA (2000) Global analysis of a model pulsing drug delivery oscillator based on chemomechanical feedback with hysteresis. Chaos 10(3):682–690

Lipinski CA, Lombardo F et al (2001) Experimental and computational approaches to estimate solubility and permeability in drug discovery and development settings. Adv Drug Deliv Rev 46(1–3):3–26

Liu JG, Nie YF (2001) Fractal scaling of effective diffusion coefficient of solute in porous media. J Environ Sci (China) 13(2):170–172

Lüdde KH, Kawakita K (1966) Die pulverkompression. Pharmazie 21:393–403

Luerkens DW (1991) Theory and application of morphological analysis: fine particles and surfaces. CRC Press, Boca Raton

Macheras P (1996) A fractal approach to heterogeneous drug distribution: calcium pharmacokinetics. Pharm Res 13(5):663–670

Macheras P, Argyrakis P (1997) Gastrointestinal drug absorption: is it time to consider heterogeneity as well as homogeneity? Pharm Res 14(7):842–847

Manai G, Delogu F et al (2002) Onset of chaotic dynamics in a ball mill: attractors merging and crisis induced intermittency. Chaos 12(3):601–609

Marsh RE, Riauka TA (2007) Modeling fractal-like drug elimination kinetics using an interacting random-walk model. Phys Rev E Stat Nonlin Soft Matter Phys 75(3 pt 1):031902

Marsh RE, Tuszynski JA (2006) Fractal Michaelis-Menten kinetics under steady state conditions: Application to mibefradil. Pharm Res 23(12):2760–2767

Martin-Landrove M, Pereira D et al (2007) Fractal analysis of tumoral lesions in brain. Conf Proc IEEE Eng Med Biol Soc 2007:1306–1309

Mihranyan A, Strømme M (2007) Solubility of fractal nanoparticles. Surf Sci 601(2):315–319

Montgomery MJ, Beringer PM et al (2001) Population pharmacokinetics and use of Monte Carlo simulation to evaluate currently recommended dosing regimens of ciprofloxacin in adult patients with cystic fibrosis. Antimicrob Agents Chemother 45(12):3468–3473

Moon SJ, Swift JB et al (2004) Role of friction in pattern formation in oscillated granular layers. Phys Rev E 69(3):031301

Muzzio FJ, Goodridge CL et al (2003) Sampling and characterization of pharmaceutical powders and granular blends. Int J Pharm 250(1):51–64

Narambuena CF, Ausar FS et al (2005) Aggregation of casein micelles by interactions with chitosans: a study by Monte Carlo simulations. J Agric Food Chem 53(2):459–463

Ottino JM, Khakhar DV (2002) Open problems in active chaotic flows: competition between chaos and order in granular materials. Chaos 12(2):400–407

Pang KS, Weiss M et al (2007) Advanced pharmacokinetic models based on organ clearance, circulatory, and fractal concepts. AAPS J 9(2):E268–E283

Papadopoulou V, Kosmidis K et al (2006) On the use of the Weibull function for the discernment of drug release mechanisms. Int J Pharm 309(1–2):44–50

Peppas NA (1985) Analysis of Fickian and non-Fickian drug release from polymers. Pharm Acta Helv 60(4):110–111

Pereira LM (2010) Fractal pharmacokinetics. Comput Math Methods Med 11(2):161–184

Persson EM, Gustafsson AS et al (2005) The effects of food on the dissolution of poorly soluble drugs in human and in model small intestinal fluids. Pharm Res 22(12):2141–2151

Rawlings JB, Miller SM et al (1993) Model identification and control of solution crystallization processes: a review. Ind Eng Chem Res 32(7):1275–1296

Roncaglia R, Mannella R et al (1994) Fractal properties of ion channels and diffusion. Math Biosci 123(1):77–101

Rowe RC, York P et al (2005) The influence of pellet shape, size and distribution on capsule filling – a preliminary evaluation of three-dimensional computer simulation using a Monte-Carlo technique. Int J Pharm 300(1–2):32–37

Sanz E, Marenduzzo D (2010) Dynamic Monte Carlo versus Brownian dynamics: a comparison for self-diffusion and crystallization in colloidal fluids. J Chem Phys 132(19):194102

Schroder M, Kleinebudde P (1995) Structure of disintegrating pellets with regard to fractal geometry. Pharm Res 12(11):1694–1700

Shah KR, Badawy SI et al (2007) Assessment of segregation potential of powder blends. Pharm Dev Technol 12(5):457–462

Shinbrot T, Alexander A et al (1999) Chaotic granular mixing. Chaos 9(3):611–620

Siepmann J, Peppas NA (2001) Modeling of drug release from delivery systems based on hydroxypropyl methylcellulose (HPMC). Adv Drug Deliv Rev 48(2–3):139–157

Sjoberg B, Mortensen K (1997) Structure and thermodynamics of nonideal solutions of colloidal particles: investigation of salt-free solutions of human serum albumin by using small-angle neutron scattering and Monte Carlo simulation. Biophys Chem 65(1):75–83

Tromelin A, Gnanou JC et al (1996) Study of morphology of reactive dissolution interface using fractal geometry. J Pharm Sci 85(9):924–928

Tromelin A, Hautbout G et al (2001) Application of fractal geometry to dissolution kinetic study of a sweetener excipient. Int J Pharm 224(1–2):131–140

Venables HJ, Wells JI (2001) Powder mixing. Drug Dev Ind Pharm 27(7):599–612

Walker DM (1966) An approximate theory for pressures and arching in hoppers. Chem Eng Sci 21:975–997

Wang TY, Sheng YJ et al (2009) Donnan potential of dilute colloidal dispersions: Monte Carlo simulations. J Colloid Interface Sci 340(2):192–201

Weidler PG, Degovics G et al (1998) Surface roughness created by acidic dissolution of synthetic goethite monitored with SAXS and N2-adsorption isotherms. J Colloid Interface Sci 197(1):1–8

Weisstein EW (2010) Fractal. http://mathworld.wolfram.com/Fractal.html. Accessed 18 Aug 2010

Xie L, Wu H et al (2008) Quality-by-design (QbD): effects of testing parameters and formulation variables on the segregation tendency of pharmaceutical powder measured by the ASTM D 6940-04 segregation tester. J Pharm Sci 97(10):4485–4497

Zook JM, Iftekharuddin KM (2005) Statistical analysis of fractal-based brain tumor detection algorithms. Magn Reson Imaging 23(5):671–678

Chapter 4
Considerations in Monitoring and Controlling Pharmaceutical Manufacturing

Pharmaceutical manufacturing is by definition a complex activity and is perhaps one of the most thoroughly studied activities from this perspective in product development (Hickey and Ganderton 2010).

Statistics and Experimental Design

Cochran and Cox (1957) were among the first to describe statistical approaches to experimental design. By implementing sound statistical principles, products with desired attributes can be prepared efficiently and reproducibly.

Sources of Error

The sources of experimental error are varied, but can be considered and ultimately controlled with appropriate experimental design and subsequent data analysis. Randomization provides a basis for experimental design that overcomes coincidental effects and allows causation to be inferred. Effects are frequently complex and do not conform to linearity or simple additive interpretation. Selected experimental designs allow for interactive and nonlinear effects to be identified without confounding with experimental error.

A.J. Hickey and H.D.C. Smyth, *Pharmaco-Complexity*, Outlines in
Pharmaceutical Sciences 1, DOI 10.1007/978-1-4419-7856-1_4,
© American Association of Pharmaceutical Scientists 2011

Randomized and Latin Square Designs

Assigning treatments to units within the design randomly is the simplest way to map data for analysis. This approach affords flexibility, as any number of replicates and treatments may be employed, and statistical ease of analysis even if replicates for some units or whole treatments are lost. The loss of information from missing data is lesser than that in other designs. However, this approach may lead to loss of accuracy resulting from uniform distribution of the whole variation across treatments and units, and becomes part of the experimental error. Different designs may be employed to reduce the error. For example, randomized blocks may be useful. Initially the design calls for dividing the experimental product into groups, which represent a single trial or replication. For example, product could be blocked for sampling time to assign specific error to environmental conditions.

Latin Square designs group treatments in replicates. The design assigns treatments to rows or columns. This eliminates errors from differences among rows and columns. Consequently, the Latin Square design reduces error compared with random blocks.

Factorial Design

Multiple variables can be evaluated concurrently by adopting a factorial design approach. The simplest approach evaluates each factor at two levels (2^n Factorial design). For example, a simple jet milling process requires consideration of two opposing gas pressures and time. Studying each factor at a low and high level, results in a 2^3 Factorial design.

Factorial experiments frequently investigate the effects of each factor over some predesignated range encompassed by the levels of that factor and are not intended to identify the combination of factors yielding the minimum or maximum response. Where factors are independent statistical analysis is easy. Additional information may be gained through confounding analysis if it is believed that factors are not independent. Factorial analysis is a rapid and efficient method for identifying an operating process space that is likely to allow product quality and performance attributes to be achieved.

Fractional Factorial Design

Full factorial designs for complex processes may be too costly and time consuming. In addition, the precision obtained may be far beyond that required for decision making. In a 2^7 factorial design, each main effect is an average of 64 combinations of other factors. It may be sufficient to conduct an 8- or 16-fold replication. Information in this partial approach is lost, in particular, with respect to interactions between factors, and so some caution should be exercised. A screening study may benefit from this approach as it has little impact, but a foundational study on which serious decisions would be based might require greater attention.

Central Composite Design

The previous designs considered linear functions between factors. Where nonlinear functions are considered, the simplest approach is a quadratic response surface (second order) obtained from central composite designs (CCDs) based on factorial analysis. CCDs test additional factors and their combinations, and can be fitted sequentially into the program of experimentation. Starting with an exploratory 2^n factorial design where the center of the first experiment is close to a point of maximum response, combinations of factors may be picked orthogonally to indicate the curvature of the response surface.

Response Surface Maps

CCDs can be extended to a broad range of combinations of factors and levels to obtain a continuous nonlinear surface that predicts the response to factor variation. For the CCD example, the response surface assessment began hypothetically when the process was near the optimum. Beyond this point, it would be desirable to approach the true response function at a small defined region around the optimum. It is important to acknowledge that the true response is curved at this location. Sequential experiments should be performed within a region of variable space, known as the operability region (OR). These experiments allow mapping over a particular region of interest, response optimization, and selection of operating conditions required to meet the specifications.

Process Design and Control

Quality by Design

A rational strategy of process or response evaluation is essential to an understanding and to the creation of knowledge. The efficiency and reproducibility of engineering processes are a challenge to the robustness of experimental designs and their ability to probe and optimize operating conditions due to the overall level of complexity.

Quality by design encompasses a range of techniques to address proactively all parameters scientifically to mitigate the risk of not meeting product quality and performance requirements, and considers the design activity from conception to market.

Tools are available to guide the product development team deliberations and include a variety of branched factor relationship diagram including, for example, the fishbone (Ishikawa) diagram. This diagram is intended to facilitate the analytical process for the group.

A fishbone diagram is illustrated in Fig. 4.1. Variables are considered in terms of the way in which they impinge on a process as assessed by a predetermined output parameter. Regardless of the stage of assessment and the purpose of the exercise, a thorough preliminary understanding of the process under consideration should result from this approach.

A thorough assessment of the process before entering into experimentation increases the potential of identifying all important variables, instills confidence that all factors have been considered,

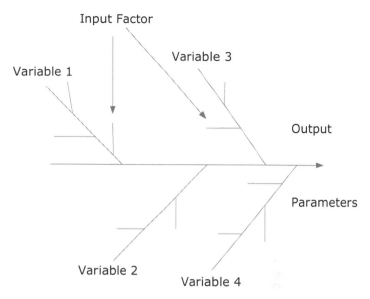

Fig. 4.1 Ishikawa (fishbone) approach to capturing input factors and relationship to output parameters (modified from Hickey and Ganderton 2010)

through the breadth of expertise minimizes potential to overlook factors, reduces time and cost of experimental missteps, and ultimately allows rational design of process space. Appropriate statistical methods as described earlier are required ideally with the ability for real-time monitoring and control through process analytical technology.

Design Space

Single response surface map and actual process design space may be distinguished by the latter being dynamic, beginning when the drug is conceived and evolving through the product lifecycle (Lapore and Spavins 2008) (Fig 4.2).

The mathematical principles of the statistical approach to experimental design would be hard to do justice in brief, but fortunately there are excellent texts on this topic (Box et al. 1978).

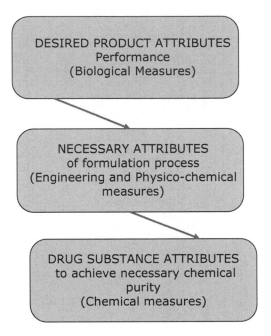

Fig. 4.2 Design space development (modified from Hickey and Ganderton 2010)

Process Analytical Technology

To exert control over any process in pharmaceutical product development, its response to changes in manufacturing variables must be monitored. Where batch production methods were employed, batch sampling followed by assessment of particular properties (drug content, particle size, etc.) was conducted. The production conditions were controlled to achieve designated quality specifications.

Quantitative analytical methods and their use have improved to allow real-time in-process measurements to be taken, and through feedback control systems, allow adjustments in input parameters to achieve continuous monitoring and control of processes.

A detailed discussion of process analytical technology (PAT) can be found elsewhere (Hickey and Ganderton 2010). A range of processes have been studied with respect to applying the principles of quality by design in the arena of engineering such as crystallization, drying, and milling.

Risk Assessment and Management

The concept of criticality is central to risk assessment and management. There are three designations with regard to criticality: noncritical, critical, and an undesignated low-risk category. Variables that have been shown not to have an impact on safety or efficacy, or factor into Critical Quality Attributes (CQA) as defined by ICH Q(8) R are considered noncritical. Consequently, they do not have to be included in design space. Critical variables are those that are known to impact safety, efficacy, or other measures of biological disposition or compliance. Critical process parameters, if varied outside a particular range, directly and significantly influence CQAs. These properties must be controlled within a predefined range and are thought to ensure final product quality. The intermediate undesignated category with regard to criticality represents attributes that may impact the product but represent a low risk. The term "low risk" is based on an indirect impact on safety and/or efficacy alone or in combination with other variables; mitigated risk; knowledge transfer from noncritical variables requires additional evaluation.

Criticality may be reduced to the elements of severity, occurrence, and detection in a compounding manner (Nosal and Schultz 2008), which in turn can be related to experimental design (frequency and variation) and analytical capability (detection). Clear differentiation of levels of criticality is required during the lifecycle of the product to facilitate a control strategy for process variables, material attributes, and their contribution to quality measures.

Given the complex nature of pharmaceutical processes and manufacturing in general, there have been attempts to adopt methods in mathematical complexity to assist in more efficiently and reproducibly guaranteeing the product quality and systems' performance. Given the likelihood that all variables are not known to the investigator, the approach that has been most common does not invoke a model that definitively fits the conditions, but rather assumes that it can be educated by knowledge of experimental results. The application of artificial neural networks (ANNs) has been addressed in detail elsewhere in the context of pharmaceutical research (Agatanovic-Kustrin and Berseford 2000), sciences (Ashanta et al. 1995), formulation (Takayama et al. 1999), and development (Borquin 1997). It suffices to say that to this point, these approaches have not superseded those of quality by design supported by statistical method, but it is to be hoped that more work on this subject will occur in coming years.

Summary

The complexity of pharmaceutical manufacturing is a challenge to control. However, unlike many other aspects of pharmaceutical product development discussed in other sections, the problem has been so severe as to warrant significant research, leading to practical approaches that support industry and regulatory standards. The basic principles behind these controls are statistically designed experiments to support definition of design space with superimposed designations for critical parameters that impinge on the quality and performance of the drug. Forays into the application of mathematical complexity in this field have begun but are limited at this point to research exercises, and will hopefully be extended to practical application in the future.

References

Agatanovic-Kustrin S, Berseford R (2000) Basic concepts of artificial neural network (ANN) modeling and its application to pharmaceutical research. J Pharm Biomed Anal 22:717–727

Ashanta AS, Kowalski JG, Rhodes CT (1995) Artificial neural networks: implications for pharmaceutical sciences. Drug Dev Ind Pharm 21:119–155

Borquin J (1997) Basic concepts of artificial neural networks (ANN) modeling in the application to pharmaceutical development. Pharm Dev Technol 2:95–109

Box GEP, Hunter WG, Hunter JS (1978) Statistics for experimenters: an introduction to design, data analysis, and model building. Wiley, New York

Cochran WG, Cox GM (1957) Experimental designs, 2nd edn. Wiley, New York

Hickey AJ, Ganderton D (2010) Pharmaceutical process engineering, 2nd edn. Informa Healthcare, New York

Lapore J, Spavins J (2008) PQLI design space. J Pharm Innov 3:79–87

Nosal R, Schultz T (2008) PQLI definition of criticality. J Pharm Innov 3:69–78

Takayama K, Fujikawa M, Nagai T (1999) Artificial neural network (ANN) as a novel method to optimize pharmaceutical formulation. Pharm Res 16:1–6

Chapter 5
Pharmacokinetics
and Pharmacodynamics

Many physiological systems appear to present data that is random or without order. The origins of this disorder are often attributed to variability introduced by the multifactorial determinants of the system. This classical view of physiological randomness has also been widely discussed in quantitative pharmacological systems. However, since the early 1990s it has been revealed that highly variable data from physiological, pharmacokinetic, and pharmacodynamic studies, in contrast to errors in measurement, have their origins in nonlinear dynamical systems that can be described by chaos theory (Goldberger 1989, 1996; Goldberger et al. 1990; Tallarida 1990a, b; van Rossum and de Bie 1991; Dokoumetzidis et al. 2001, 2002; Mager and Abernethy 2007).

Pharmacokinetics/Toxicokinetics

The interaction of drugs and exogenous substances with the body, in particular, the time course of these compounds within the body, falls in the general field of pharmacokinetics (i.e., "what the body does to the drug"). Absorption, distribution, metabolism, and excretion are all processes that occur sequentially and simultaneously upon administration of a drug/toxicant to the organism. The pharmacokineticist must assess these processes via sampling of body fluids

A.J. Hickey and H.D.C. Smyth, *Pharmaco-Complexity*, Outlines in Pharmaceutical Sciences 1, DOI 10.1007/978-1-4419-7856-1_5, © American Association of Pharmaceutical Scientists 2011

and use of pharmacokinetic models. There is often a relationship between plasma drug concentrations and pharmacologic responses but these are also often complex. Interpretation of pharmacokinetic data involves mathematical modeling to enable prediction of blood/plasma and tissue concentrations time profiles, and facilitate mechanistic understanding of the underlying processes. It is widely recognized that these models currently used are vast oversimplifications "but we have to start somewhere" (Pang et al. 2007).

Pharmacokinetics must predict the safety and efficacy of drugs and is a major component of the development and regulation of pharmaceuticals. Drug developers must now use pharmacokinetics to determine the optimal dose where the most benefit is obtained with the least risk of potential side effects. Simultaneously, with these increasing regulatory pressures, some have made the observation that around 95% of new drugs have suboptimal pharmacokinetic profiles and changes "a possible blockbuster into just another has-been" (Orive et al. 2004). Drug discovery via high-throughput screening has also tended to produce candidate drugs with physicochemical properties that compromise pharmacokinetic profiles. Although pharmacokinetics is seen as a critical for its predictive power, the tools widely employed to enable optimal prediction may not represent reality closely enough. Moreover, it might be imagined that given suboptimal or complex pharmacokinetic profiles, delivery systems may be designed, via a more comprehensive understanding of the underlying pharmacokinetic mechanisms, to maximize the drug benefit while minimizing risk.

Classical PK is reduced to linear relationships but actually is modeled by exponential (log) functions. It is a phenomenon very similar to chemical kinetics in that it can be modeled by linearity, but the underlying complexity which is more informative is lost in the averaging phenomenon (Sect. 2).

Compartmental Modeling

Of the various approaches used to predict drug time profiles in the body, compartmental models are the simplest and most widely used (Pang et al. 2007). The basis of the compartment model is equilibrium systems, homogeneity, and mass transfer between compartments. A compartment is defined as the number of drug molecules having the same probability of undergoing a set of chemical kinetic processes (Marsh and Tuszynski 2006). The mass transfer of drug molecules between compartments is described by kinetic rate constants. As alluded to above, these models assume that each compartment is homogenous (instantaneous mixing) and mass transfer constants are constant. Therefore, the system can be described by coupled first-order differential equations. Macheras and colleagues argue that the assumptions of homogeneity and well-stirred media are not supported by anatomical and physiological evidence (Macheras et al. 1996; Macheras and Argyrakis 1997; Dokoumetzidis et al. 2004; Pang et al. 2007).

Fractal Kinetics

The realization that physiological and anatomical characteristics of the human body are complex and not well described by compartment models has lead several researchers to investigate pharmacokinetics from a fractal point of view. The concept of fractals was introduced in Sects. 2 and 3. The argument made is that diffusion, a major component of drug transport and kinetics, is not well described by Fick's laws in the human body because of under-stirred areas and constrained spaces are known to exist. Furthermore, biological systems are made up of a multitude of interacting parts and will be nonlinear from a dynamical systems point of view (Dokoumetzidis et al. 2004). As mentioned earlier in this text, fractals have the property of being self-similar at various scales of scrutiny. In physiology, fractals can be found in numerous organ systems, for example, the vascular tree, the lungs, and the folds of the brain have all been well described using fractal geometry. Application of fractals to pharmacokinetics has been used for drug release and dissolution (as shown in Sect. 3), absorption (Macheras and Argyrakis 1997), and fractal compartment models (Fuite et al. 2002; Marsh and Tuszynski 2006). Excellent reviews of these fractal concepts applied to pharmacokinetics have been published (Dokoumetzidis et al. 2004; Pang et al. 2007).

Physiologically Relevant Modeling

Physiologically based pharmacokinetic (PBPK) models relate organ or tissue structures to the physiology of the organ or tissue. The PBPK models can be considered as a data-informed approach to achieve a more biologically realistic dose–response model. PBPK models facilitate the estimation of drug concentrations at target and off-target tissues by taking into account the rate of absorption into the body, distribution, and storage in tissues, metabolism, and excretion on the basis of interplay among critical physiological, physicochemical, and biochemical determinants. Qualitative evidence that this type of modeling is more relevant can be observed from those agencies and institutions involved in risk assessment of exposure. For example, physiologically based pharmacokinetic models have increasingly been employed in chemical health risk assessments carried out by the US Environmental Protection Agency (EPA), and it is anticipated that their use will continue to increase. Relevant physiological parameter values (e.g., alveolar ventilation, blood flow and tissue volumes, and glomerular filtration rate) are critical components of these models, and values can now be found in various databases for use by researchers and risk assessors. From a mathematical point of view, a PBPK model comprises a system of coupled ordinary differential equations (ODEs). These equations involve physiological and physicochemical parameters, each of which is typically affected with uncertainty and some degree of variability due to inter- and intra-individual variations. To investigate the effects of initial values and uncertainty in parameters for these models (and the ODEs), Monte Carlo methods have been used (Thomas et al. 1996).

Efficacy/Safety

In linear systems the result of an input, such as the efficacy or toxicity of a drug, is proportional to the stimulus (e.g., dose) and multiple stimuli result in the summation of the inputs. These systems are quite attractive to those wishing to understand and predict the effects of stimuli (i.e., responses to a certain dose) as classical mathematics and models that can describe this type of linearity are very familiar to us. However, in nonlinear systems, the resulting behavior of a series of inputs is not equal to the summation of all the components and individual behaviors. These basic principles of complex systems and chaos have now been widely reported and popularized. However, application of these methods of interpretation and modeling of pharmacokinetic and pharmacodynamic data is still relatively rare. This, no doubt, is due to the unfamiliarity of the mathematical methods and the absence of standard training in this field for biologists and pharmaceutical scientists.

It is clear, however, physiological systems are nonlinear. The advantage of nonlinear systems for physiological processes is suggested to be one of energy efficiency and control compared to linear systems (Dokoumetzidis et al. 2001). To predict the future state of a dynamical system requires iterating the mathematical function many times, advancing time in small steps. This iteration procedure is referred to as "solving the system" or integrating. Using fast computer processing now readily available, it is possible to determine all future points using just the initial point. This time series is known as the trajectory or orbit. In nonlinear dynamical systems, the data appear to be random and seemingly unpredictable and has been described as chaos. The mathematics related to nonlinear dynamical systems and chaos focuses not on "solving the system" and finding precise solutions to the governing equations (as this often appears hopeless), but rather on determining qualitatively how the system will behave in the long term. This is because chaotic systems are very sensitive to small differences in initial conditions, i.e., errors due to very small rounding errors in numerical computation may result in widely diverging outcomes. Thus, despite the fact that chaotic systems are deterministic (their future of the system is fully determined by the initial conditions and not random behavior), the high sensitivity of the system on initial conditions makes long-term prediction of outcomes very difficult and often impossible. In summary, chaotic systems are deterministic but not predictable. A classical example is the weather (Smith 2002) but chaos can also be found within trivial systems, and therefore it is not surprising that physiological responses to drug administration may also be described by nonlinear dynamics and chaos.

Classical pharmacodynamic models and approaches are unable to adequately deal with such complex and variable data. Ligand–receptor interactions, upon which traditional pharmacodynamics is largely based have been shown to be complex (Prank et al. 1995). The classical Emax model widely used in pharmacodynamic studies (Schoemaker et al. 1998) may not sufficiently deal with deviations that are commonly observed in practice, such as feedback mechanisms induced by ligand–receptor interactions that nonlinearly influence the pharmacodynamic response (Dokoumetzidis et al. 2001). In fact, many examples are readily available from closely related fields:

(a) Physiological responses to hormones and neurotransmitters: these endogenous substances interact with cellular receptors in controlled reactions governed by the mass-action law and some subsets of receptors result in negative feedback that reduces further release rates of the ligand (Tallarida and Freeman 1994). The system was shown to pass from periodic to chaotic as the input parameters varied.

(b) Cytokines, small protein or glycoprotein messenger molecules, allow transfer of information from one cell to another. There are large numbers of different types of cytokines that can trigger complex intracellular signaling cascades in other cells. Individual cytokines have multiple and diverse biological functions. A panoply of feedback loops operates within these complex cytokine–cellular systems that results in complex nonlinear behavior (Callard et al. 1999; Seely and Christou 2000; Higgins 2002).

Additional examples are found in an excellent review by Dokoumetzidis et al. (2001).

Mager and Abernethy (2007) cite that the biological variability in biological signaling and the complexity of pharmacological systems often complicate or preclude the direct application of traditional structural and nonstructural models in pharmacodynamics. They suggest that mathematical transforms (such as fast Fourier and wavelet transforms) of data sets may be used to provide better measures of drug effects and patterns in responses, and assist in interpretation of diverse data sets (e.g., imaging and biomedical signals). As observed in our earlier example of powder flow analysis, mathematical transforms may also be required for pharmacodynamic data interpretation as the time-series data is nonlinear and the frequency content of a signal is more informative than the original waveform.

Summary

Methods to deal with complexity in PK/PD studies is key to establishing how drugs should be administered is a risk analysis task, achieved by defining and analyzing the potential benefits and dangers to individuals and populations. In practice, new chemical entities are evaluated in vitro, in preclinical models in vivo, then in human using clinical trials. Early trials are focused on safety and establishing potential risks of the drug. The larger scale clinical trials focus on efficacy but also must gather data on safety. These studies are analyzed statistically and approval is generally based on envisioning the average patient. However, it is clear that current methods may not be able to predict or guide drug development at these later phases to the extent desired.

References

Callard R, George AJ et al (1999) Cytokines, chaos, and complexity. Immunity 11(5):507–513

Dokoumetzidis A, Iliadis A et al (2001) Nonlinear dynamics and chaos theory: concepts and applications relevant to pharmacodynamics. Pharm Res 18(4):415–426

Dokoumetzidis A, Iliadis A et al (2002) Nonlinear dynamics in clinical pharmacology: the paradigm of cortisol secretion and suppression. Br J Clin Pharmacol 54(1):21–29

Dokoumetzidis A, Karalis V et al (2004) The heterogeneous course of drug transit through the body. Trends Pharmacol Sci 25(3):140–146

Fuite J, Marsh R et al (2002) Fractal pharmacokinetics of the drug mibefradil in the liver. Phys Rev E Stat Nonlin Soft Matter Phys 66(2 pt 1):021904

Goldberger AL (1989) Cardiac chaos. Science 243(4897):1419

Goldberger AL (1996) Non-linear dynamics for clinicians: chaos theory, fractals, and complexity at the bedside. Lancet 347(9011):1312–1314

Goldberger AL, Rigney DR et al (1990) Chaos and fractals in human physiology. Sci Am 262(2):42–49

Higgins JP (2002) Nonlinear systems in medicine. Yale J Biol Med 75(5–6):247–260

Macheras P, Argyrakis P (1997) Gastrointestinal drug absorption: is it time to consider heterogeneity as well as homogeneity? Pharm Res 14(7):842–847

Macheras P, Argyrakis P et al (1996) Fractal geometry, fractal kinetics and chaos en route to biopharmaceutical sciences. Eur J Drug Metab Pharmacokinet 21(2):77–86

Mager DE, Abernethy DR (2007) Use of wavelet and fast Fourier transforms in pharmacodynamics. J Pharmacol Exp Ther 321(2):423–430

Marsh RE, Tuszynski JA (2006) Fractal Michaelis-Menten kinetics under steady state conditions: application to mibefradil. Pharm Res 23(12):2760–2767

Orive G, Gascon AR et al (2004) Techniques: new approaches to the delivery of biopharmaceuticals. Trends Pharmacol Sci 25(7):382–387

Pang KS, Weiss M et al (2007) Advanced pharmacokinetic models based on organ clearance, circulatory, and fractal concepts. AAPS J 9(2):E268–E283

Prank K, Harms H et al (1995) Nonlinear dynamics in pulsatile secretion of parathyroid hormone in normal human subjects. Chaos 5(1):76–81

Schoemaker RC, van Gerven JM et al (1998) Estimating potency for the Emax-model without attaining maximal effects. J Pharmacokinet Biopharm 26(5):581–593

Seely AJ, Christou NV (2000) Multiple organ dysfunction syndrome: exploring the paradigm of complex nonlinear systems. Crit Care Med 28(7):2193–2200

Smith LA (2002) What might we learn from climate forecasts? Proc Natl Acad Sci U S A 99(suppl 1):2487–2492

Tallarida RJ (1990a) Control and oscillation in ligand receptor interactions according to the law of mass action. Life Sci 46(22):1559–1568

Tallarida RJ (1990b) Further characterization of a control model for ligand–receptor interaction: phase plane geometry, stability, and oscillation. Ann Biomed Eng 18(6):671–684

Tallarida RJ, Freeman KA (1994) Chaos and control in mass-action binding of endogenous compounds. Ann Biomed Eng 22(2):153–161

Thomas RS, Bigelow PL et al (1996) Variability in biological exposure indices using physiologically based pharmacokinetic modeling and Monte Carlo simulation. Am Ind Hyg Assoc J 57(1):23–32

van Rossum JM, de Bie JE (1991) Chaos and illusion. Trends Pharmacol Sci 12(10):379–383

Chapter 6
Impact of Complexity on Population Biology

The previous chapters have demonstrated that physical chemistry, engineering, and consequently the regulatory considerations are required to develop pharmaceutical products; when scrutinized closely it involves complex and confounding variables that require a high level of control.

Clinically it is recognized that human organisms and their homeostatic mechanisms are also very complex, but historically little has been done to recognize this in approaches to therapy. As novel discoveries in the biological sciences and biotechnology are brought to bear on disease, revolutionary new treatments are likely to be developed.

Nature of Disease

Populations of people are diverse with respect to phenotype in terms of age, gender, ethnicity, and also geographical location (important for environmental stimuli/exposure, diet, etc.); also genotype which underpins the phenotype and results in different responses to aberrant cell/tissue types or pathogens. The additional layer of complexity related to the proteome and metabolomics, or regulation of gene expression and metabolism renders the biological system very complex. Consequently, the predictive

A.J. Hickey and H.D.C. Smyth, *Pharmaco-Complexity*, Outlines in Pharmaceutical Sciences 1, DOI 10.1007/978-1-4419-7856-1_6, © American Association of Pharmaceutical Scientists 2011

nature of treatment or prevention of disease is subject to enormous variability.

At the beginning of the millennium, the first publications on whole human genome sequences opened the possibility of revolutionary approaches to disease treatment and prevention (Venter et al. 2001; Levy et al. 2007). Initially the pharmaceutical industry believed that mapping single nucleotide polymorphisms would indicate important new targets for drug therapy (Altshuler et al. 2000). However, as rare diseases become a focus for new therapeutic approaches, compilations of individual genomes of affected individuals seem to be more likely to result in identification of targets (Turnbull et al. 2010). The complex biology that results in translation and expression of genes into major functional building block proteins makes the proteome an important and diverse component of the interaction of an organism with its environment (O'Donovan et al. 2001). Indeed, since functionality is the key to understanding, prediction and control of disease, a large emphasis has been placed on a sub-category of proteins that are involved in metabolomics (Pearson 2007).

Potential for Disease Intervention

There has been a considerable interest in mathematical approaches to the population biology of infectious diseases for many years (Anderson and May 1979; May and Anderson 1979). The ability to rapidly sequence infectious micro-organisms, such as *Mycobacterium tuberculosis* (Cole et al. 1998), put researchers in a unique position to consider bioinformatic approaches to map genome, proteome, and metabolome of infectious micro-organisms as therapeutic targets.

In the first instance the outcome of these new discoveries and the evolution of a longstanding interest in population biology are targeted therapeutic strategies for specific diseases, such as cancer (Gatenby and Vincent 2003). However, the logical outcome of the developments in molecular and cellular biology is the possibility of individualized therapies and preventative strategies (Evans and Relling 2004; Hirano 2007).

The demand for more unique approaches to therapy will feedback to the properties of drugs, dosage forms, and their manufacture, requiring much greater flexibility and speed of product development. This will only be achieved by maximizing the data collection (as is already occurring in molecular biology) and subjecting it to rigorous scrutiny to yield as much interpretable information and knowledge as possible to facilitate decision making and efficient pharmaceutical project and risk management (Kennedy 1998; Vesper 2006).

Epidemiological Studies and Adverse Events

Epidemiological studies investigating causal hypotheses often are inconsistent from one study to the next depending on the methodology. Contradictory results may be explained through complexity and nonlinearity (Glattre and Nygard 2004). The possibly fractal nature of ordered series of relative risks (RR) and their possible self-organized criticality (SOC) have been suggested. Using these concepts, Glattre and Nygard performed reanalysis of published meta-studies, one of which investigates the possible association of oral contraceptives and female breast cancer, and found different conclusions than those made by the original study.

The incidence of adverse drug reactions are typically reported in a statistical method but a few studies have, from a population stand-point, investigated the variability of responses or the root cause of why they occur. Frattarelli showed, with limitations, that the severity of adverse drug reactions follows a distribution seen in other complex adaptive systems, called a power law distribution, and that preventable reactions occurred for reasons fundamentally different from those that underlie the nonpreventable reactions (Frattarelli 2005; Dokoumetzidis and Macheras 2006). Using published studies, the author plotted incidence of drug reaction as a function of severity and then performed a fit to an equation. Overall and nonpreventable drug reactions followed a power law distribution regardless of sample size or the nature of the population or drugs studied. Preventable reactions on the other hand, were described by a different type of equation.

Summary

The enormous strides that have been made in understanding human and pathogen genomes, proteomes, and metabolomes give the medical and pharmaceutical community a unique opportunity in the history of mankind's pursuit of knowledge to truly develop targeted therapies to address populations and subpopulations of people for variants on particular diseases. Moreover, taken to its logical conclusion the ability to differentiate the particular manifestations of a disease associated with an individual's therapies, tailored to their specific needs, become a distinct possibility. Having followed the theme of this small volume it should be evident that such a development would put enormous pressure on pharmaceutical scientists and engineers to adapt to a rapidly changing product development and manufacturing environment, to bring all of the relevant monitoring and controls of the complex phenomena described earlier, to bear, and to support this radical new approach to disease therapy and prevention.

References

Altshuler D, Pollara VJ, Cowles CR, Van Etten WJ, Baldwin J et al (2000) An SNP map of the human genome generated by reduced representation shotgun sequencing. Nature 407:513–516

Anderson RM, May RM (1979) Population biology of infectious disease: Part I. Nature 280:361–367

Cole ST, Brosch R, Parkhill R, Garnier T, Churcher C et al (1998) Deciphering the biology of *Mycobacterium tuberculosis* from the complete genome sequence. Nature 393:537–544

Dokoumetzidis A, Macheras P (2006) A comment on "Adverse drug reactions and avalanches: life at the edge of chaos". J Clin Pharmacol 46(9):1057–1058, author reply 1059–1060

Evans WE, Relling MV (2004) Moving toward individualized medicine with pharmacogenomics. Nature 429:464–468

Frattarelli DA (2005) Adverse drug reactions and avalanches: life at the edge of chaos. J Clin Pharmacol 45(8):866–871

Gatenby RA, Vincent TL (2003) Application of quantitative models from population biology and evolutionary game theory to tumor therapeutic strategies. Mol Cancer Ther 2:919–927

Glattre E, Nygard JF (2004) Fractal meta-analysis and 'causality' embedded in complexity: advanced understanding of disease etiology. Nonlinear Dynamics Psychol Life Sci 8(3):315–344

Hirano T (2007) Cellular pharmacodynamics of immunosuppressive drugs for individualized medicine. Int Immunopharmacol 7:3–22

Kennedy T (1998) Pharmaceutical project management. Dekker, New York

Levy S, Sutton G, Ng PC, Feuk L, Halpern AL et al (2007) The diploid genome sequence of an individual human. PLoS Biol 5(10):e254. doi:10.1371/journal.pbio 0050254

May RM, Anderson RM (1979) Population biology of infectious disease: Part II. Nature 280:455–461

O'Donovan C, Apweiler R, Bairoch A (2001) The human proteomics initiative (HPI). Trends Biotechnol 19:178–181

Pearson H (2007) Meet the human metabolome. Nature 446:8. doi:101038/446008a

Turnbull C, Ahmed S, Morrison J, Pernet D, Renwick A et al (2010) Genome-wide association study identifies five new breast cancer susceptible loci. Nat Genet. doi:10.1038/ng.586

Venter JC, Adams MD, Myers EW, Li PW, Muraj RJ et al (2001) The sequence of the human genome. Science 291:1304–1351

Vesper JL (2006) Risk assessment and risk management in the pharmaceutical industry: clear and simple. Parenteral Drug Association, Washington

Chapter 7
Conclusion

An all-too-pervasive modern myth holds that science consists in the discovery and collection of authoritative facts or truths. This myth stands in contrast to the experience of all genuine scientists, that ours is really just the passionate quest for what nature has to tell us.

Victor R. Baker
University of Arizona
Times Literary Supplement
July 17th, 2009

It should be no surprise to pharmaceutical scientists and engineers that they work in an arena of complex multivariate phenomena which require costly and time-consuming efforts to monitor, understand, predict, and control. Indeed, it might be argued that an outcome of this small volume is to state the obvious. However, the approaches taken to maneuver in order to gain as much knowledge as possible if the systems involved to integrate them into larger product development and therapeutic objectives are dictated by the ability to acquire data reflecting the actual nature of the situation and interpret it appropriately.

In Chap. 1, the concepts of mathematical complexity consisting of tools to address data were introduced briefly. The more philosophical concept was raised of the way in which we communicate our findings to improve products and therapies, and the hierarchical bias that we often use to depict thinking on this topic. Moreover, the complex ways in which fields "overlap" do not fully capture,

A.J. Hickey and H.D.C. Smyth, *Pharmaco-Complexity*, Outlines in Pharmaceutical Sciences 1, DOI 10.1007/978-1-4419-7856-1_7, © American Association of Pharmaceutical Scientists 2011

from an intellectual perspective, the integrative nature of each of these elements. It was proposed that the reflections of the following chapters might lead us to a slightly different way of considering these issues.

Chapter 2 introduces the notion that complexity in physics and chemistry underpinning important pharmaceutical principles is not a new observation. However, it should be noted that the apparent simplicity of our interpretation of molecular dynamics and chemical kinetics belies a level of complexity that we have yet to fully probe and this may be of future importance. The evolution of understanding, over approximately a century, of aspects of molecular association and surface chemistry is described as a foundation for considering recent developments. It is evident as we scrutinize systems more closely that they become susceptible to a level of understanding that in the absence of current analytical and probing methods would not be possible. In short, increasing the amount of available data allows for more accurate interpretation of observations and ultimately this knowledge will facilitate prediction and control.

Chapter 3 describes the importance of complex physico-chemical properties and methods of interpreting them with respect to the performance of solid dosage forms which represent the majority of pharmaceutical products. The convergence of mathematical complexity and the true nature of these systems has been one of the intriguing occurrences in the last two decades, as it points to the potential for a level of prediction and control that has hitherto been unimaginable.

Chapter 4 describes the considerations that are uppermost in pharmaceutical manufacturing processes. This topic is of great concern to scientists, engineers, and regulators trying to bring the highest level of control to the quality and performance, thereby serving patients with reliable and therapeutic products. The mathematical approaches to this have focused predominantly on statistical process control. It might be proposed that the number of unknown variables, essential to the complete nature of the process, is not or cannot be known. Consequently, the most efficient approach, at this time, is to isolate the variables understood to contribute most significantly to the process and to mitigate against unknown environmental or time-dependent variables. However, there are fields in mathematical complexity that would allow an intelligent approach wherein the analytical system "learns" about the process and brings that knowledge to bear on control. Notably, the field of artificial neural networks has seen the greatest interest for this purpose and presumably will continue to evolve to suit the product development needs.

In order to do justice to the scope of product development, clinically important parameters of pharmacokinetics and pharmaco-dynamics were addressed in Chap. 5. It is clear that the nature of the biological response to the administration of a therapeutic (or preventative) agent is as complicated as the dosage form and its production. These topics have historically been approached with simplifying assumptions but the capacity to develop physiologically and pharmacologically relevant systems indicates the great potential to address Ehrlich's goal of the "magic bullet" (Albert 1960).

The impact of a full understanding of population biology on approaches to product development is necessarily speculative as described in Chap. 6. Since the last decade can, without hyper-bole, be considered the dawn of a new age in biology much has yet to be done. Nevertheless, our current knowledge of genomics, proteomics, and metabolomics indicates that we will need com-plex analyses as presaged by the rising interest in pharmaco-informatics. It is likely that research in biology will allow specific drug targeting within populations and subpopulations, and ulti-mately at the level of routine individualized medicine. This will undoubtedly feedback to the earlier stages of product develop-ment, requiring speed and flexibility and much greater under-standing of the processes involved in order to continue to ensure product quality and performance.

Reflecting on the conception of the way in which we manage data to gain understanding as presented in Chap. 1, and in light of the multiple complex processes involved in product development, a more general scheme can be proposed. If we assume this, there are existing domains in the realms of understanding, i.e., data, information, knowledge, and wisdom as shown in Fig. 7.1. Arguably, these might be considered infinite in that we are not aware of boundaries. Then we can map one domain onto another

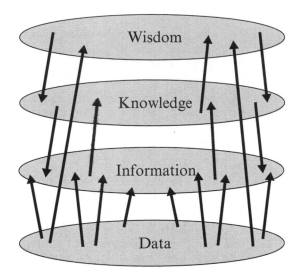

Fig. 7.1 Schematic depiction of the way in which data impinge on understanding (information, knowledge, and wisdom) indicating the proposition that the mapping occurs through domains which represent "all" of the items indicated, and consequently imply a link between all relevant phenomena that ultimately, in an ideal world, allows prediction and control of any phenomenon

and there are reflections which occur from "higher" to "lower" levels. Exploring the relationships within not only one process but interfaces or integrations into others may make progress in treatment and prevention of disease more rapid and advance our understanding of "the big picture." Undoubtedly, from a practical standpoint this can only be achieved by the application of computers, their increased processing capacity and information technology. Whether, as has been suggested elsewhere, we need to invoke sentient computers to fully explore the domains identified in Fig. 7.1 is beyond the scope of this text but is an interesting philosophical question (Tipler 1994).

Taking the general principles of knowledge management and returning to the conception of the relationship between the processes discussed in this volume, it is clear that there is a domain of all possible factors involved in product development,

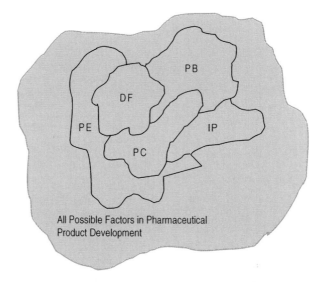

Fig. 7.2 Schematic depiction of the complex relationships between the selected processes described (*PE* pharmaceutical engineering (manufacturing); *DF* dosage form; physico-chemical factors; *IP* individual pharmacokinetics and pharmacodynamics and; *PB* population biology)

and in order to predict and control the ones we can clearly define (and hopefully others that might be identified in the future), their complex interactions must be explored to understand the continuum from drug molecule to disease treatment or prevention (Fig. 7.2).

The goal of this outline has been to present the practical implications of pharmaco-complexity in selected scientific areas of importance to pharmaceutical product development and in turn to the pharmaceutical scientist. In presenting the factual background, an attempt has been made to raise more philosophical questions regarding the way we approach these complex phenomena in order to gain the level of understanding required to guarantee the quality and performance of products. If this encourages greater interdisciplinary thinking on the nature of future therapy, this will justify the effort expended.

References

Albert A (1960) Selective toxicity: the physico-chemical basis of therapy, 2nd edn. Australian National University/Wiley, Canberra/New York

Tipler FJ (1994) The physics of immortality. Anchor Books/Doubleday, New York

Index

B
Batch production methods, 42
BET theory, 11

C
Carr's compressibility index, 27
Central composite design (CCD), 39
Classical Emax model, 52
Critical quality attributes (CQA), 43

D
Data, information, knowledge,
    and wisdom, 63, 64

H
Hartley model, 8
Hixon-Crowell cube root law, 21–22
Homeostatic mechanisms, 55

K
Knowledge management, 64

L
Langmuir's adsorption equation,
    11, 12
Latin square designs, 38

M
Mass action law, 8
Micelle, 7
Mycobacterium tuberculosis, 56

N
New chemical entities (NCE), 19
Noyes-Whitney equation, 21

P
Pharmaceutical manufacturing,
    62–63
  definition, 37
  process design and control
    batch production methods, 42
    design space development, 42
    quality, 40–41
    quantitative analytical
      methods, 43
  risk assessment and
    management, 43–44
  statistics and experimental
    design
    CCD, 39
    factorial design, 38–39
    fractional factorial design, 39
    latin square designs, 38
    response surface maps, 39–40
    sources of error, 37

About the Authors

Dr. Anthony Hickey is Professor of Molecular Pharmaceutics and Biomedical Engineering at the University of North Carolina at Chapel Hill. He is a fellow of the Institute of Biology, American Association of Pharmaceutical Scientists and the American Association for the Advancement of Science. He has published several edited and authored volumes in the fields of pharmaceutical aerosols, process engineering and particulate science.

Dr. Hugh Smyth is Assistant Professor of Pharmaceutics at the University of Texas, Austin. He is a recipient of the Young Investigator in Pharmaceutics and Pharmaceutical Technology Award of the American Association of Pharmaceutical Scientists and has edited a volume on pulmonary drug delivery. Drs. Hickey and Smyth share a research interest in the delivery of drugs to the lungs for the treatment and prevention of a number of diseases.

A.J. Hickey and H.D.C. Smyth, *Pharmaco-Complexity*, Outlines in Pharmaceutical Sciences 1, DOI 10.1007/978-1-4419-7856-1, © American Association of Pharmaceutical Scientists 2011